International Migration and Knowledge

Two unconnected but important recent academic and policy debates have focussed on the idea of the knowledge-based economy and the economic consequences of increasing international migration. This book challenges pre-conceived views on the debates and argues the need to understand that all migrants are potentially knowledge carriers and learners, and that they play an essential role in the globalization of knowledge transactions.

Deconstructing the concept of knowledge, and demonstrating how tacit knowledge is in fact an amalgam of encultured and embrained/embodied forms of knowledge, this book considers how international migration has profound consequences, analyzed, first, in terms of the economic and immigration strategies of national and regional bodies. And, second, the authors explore how the 'diversity dividend' of migration is captured by firms through their management strategies, and by individuals through increasingly boundaryless careers, continuous learning and transnational working lives.

This research is a highly original contribution which provides the first overview of one of the most dynamic forces for change in the globalizing economy. It will challenge migration researchers and students to engage with the management and learning literatures, and it will challenge management and economic policy analysts to think through the role of international migration. As such it will contribute to teaching and research in a range of social science disciplines, as well as to those involved in policy arenas ensuring that firms and all migrants engage in mutual learning and knowledge sharing.

Allan M. Williams is Professor of European Integration and Globalization at London Metropolitan University, UK.

Vladimir Baláž is a Senior Researcher in the Institute of Forecasting in Slovakia

Routledge Studies in Human Geography

This series provides a forum for innovative, vibrant, and critical debate within Human Geography. Titles will reflect the wealth of research which is taking place in this diverse and ever-expanding field. Contributions will be drawn from the main sub-disciplines and from innovative areas of work which have no particular sub-disciplinary allegiances.

International Migration and Knowledge

Allan M. Williams and Vladimir Baláž

Routledge
Taylor & Francis Group

LONDON AND NEW YORK

First published 2008 by Routledge
2 Park Square, Milton Park, Abingdon, Oxfordshire OX14 4RN

Simultaneously published in the USA and Canada
by Routledge
711 Third Avenue, New York, NY 10017

First issued in paperback 2014

*Routledge is an imprint of the Taylor & Francis Group,
an informa business*

© 2008 Allan M. Williams and Vladimír Baláž

Typeset in 10/12pt Times NR MT by Graphicraft Limited,
Hong Kong

British Library Cataloguing in Publication Data
A catalogue record for this book is available from the British
Library

Library of Congress Cataloging in Publication Data
A catalog record for this book has been requested

ISBN 978-0-415-43492-8 (hbk)
ISBN 978-0-415-76184-0 (pbk)
ISBN 978-0-203-89465-1 (ebk)

Contents

Figures

Tables

Boxes

Preface

This book has its origins in several joint projects that we have undertaken in the past decade on the relationships between mobility and economic development in the emerging economies of Central and Eastern Europe, and on returned migration to Slovakia from the UK. Three of the main themes in this research were: the importance of migration as a conduit for learning and knowledge transfer; the need to look at the full cycle, or cycles, of migration; and the potential for all migrants to be knowledge bearers and co-learners.

Drawing on this earlier research, and our separate work on the theorization of knowledge, and the nature of innovation and R&D, our aim has been to write a book which explores the relationship between international migration and knowledge. There is now widespread agreement as to the importance of tacit knowledge in economic performance and competitiveness. Our central argument is that international migration is an important, and sometimes the only, effective means for transferring some forms of tacit knowledge across international boundaries. One of the biggest obstacles to such transfers is institutional differences, for knowledge, and the use made of knowledge, is based not only on structural economic differences but also on differences in norms, practices and cultures. However, this is both an obstacle and an opportunity. Those migrants who can think and act with deep reflexivity are potential boundary spanners who can transfer knowledge across space and between economic systems. Migration is also more than just a source of substitute human capital, for it can be a source of diversity and creativity.

Of course, not all migrants have positive experiences of knowledge transfer and learning and very few constitute significant boundary spanners or knowledge brokers. It is also the case that the more highly-skilled migrants are likely to have a far greater impact on productivity and performance in receiving and destination economies, compared certainly to those who are relatively less skilled. However, we emphasize the need to look beyond the elite places, migrants and sectors that have usually figured in this research field. While it is unarguable that all migrants are involved in some forms of learning – whether of technical skills, language or culture – it is also true that the vast majority are potential sources of knowledge transfer, although this is likely to be of modest scale in most individual cases. National or regional economies which declare themselves to be learning economies need to consider how best to harvest the unusual knowledge that many migrants bring with them, and the same applies to firms. There is also a need to rethink national and regional innovation systems, which often pay scant attention to international migrants, and to see migration policies as potential knowledge transfer and innovation policies.

Although we have tried to adopt a holistic approach, this does not mean that the book is comprehensive. No volume could be in addressing this vast research arena. Instead, we have focused on exploring how changes in the nature of migration intersect with our understanding of knowledge and learning, and have sought to analyse this at a number of different levels – the national, the regional, the firm and the individual – while emphasizing that these levels are inter-related. Our examples, mostly set out in a series of boxes throughout the volume, have drawn on a wide range of settings, including sending and receiving countries, and diverse forms of migration and mobility. In practice, the boxes focus more on developed and emerging economies, than on the rest of the world, but this reflects the available literature in this field. The emerging picture is one of complex interactions between changing systems of migration and mobility, and changing economic systems. These defy easy generalization but, we believe, emphasize the increasing importance of understanding the inter-relationships between knowledge and international migration.

Finally we acknowledge the assistance of our colleagues at the Institute for Forecasting (Bratislava) and London Metropolitan University in preparing this book, and to Linda Williams and Denisa Vörösová-Balážová for their general support and encouragement. Allan Williams also acknowledges the support of a British Academy Readership for creating the time and space for developing some of the theoretical content of the book. Vlado Baláž is pleased to acknowledge the support he has received from the British Academy and the Slovak Academy during several visits to the UK.

Vladimir Baláž and Allan Williams
Bratislava and London, 2008

1 Introduction

Mobility and knowledge: global challenges

The significance of knowledge to modern economies has long been widely acknowledged (Brown *et al.* 2001; Nonaka and Takeuchi 1995), and this has evolved into more specific notions about learning and knowledge-based economies. In essence, these emphasize that knowledge is not so much a 'factor of production' but *the* key determinant of productivity, competitiveness and economic growth. Not untypical is Welch's (2001: 21; emphases added) assertion that: 'Know that the ultimate, sustainable competitive advantage lies in the ability to learn, *to transfer* that learning across components, and to act on it *quickly*'.

We return to the nature of knowledge later in this chapter, and at greater length in Chapter 3, but focus here on the words emphasized in Welch's quotation – 'transfer' and 'quickly'. Knowledge in itself is of limited value, for what matters is how that knowledge is collected, transferred and applied. This was emphasized by Peter Drucker, the management guru who has played a key role in promoting the role played by knowledge. In a seminal paper, Drucker (1993: 176) argued that 'to make knowledge you have "to learn to connect"'. It is the notion of 'connecting' that is central in this volume, or more precisely how connections can be made that allow knowledge to be used in economic activities.

There are fundamentally two ways in which a particular unit (whether a firm, a region, or a national state) can generate or enhance the knowledge that is available to be applied in economic activities: endogenously, or exogenously, that is internally or externally. With the growing interest in theorizing the role of knowledge in the economy, in recent decades, attention initially focused on endogenous sources. This took several forms, including the debates about national innovation systems, stimulating Research and Development (R&D) capacity and investment in education, skills and training. In addition, economic geographers and regional economists, in particular, focused on one particular aspect of knowledge transfer, that is the importance of spatial proximity in the transfer of tacit knowledge via face-to-face contacts, particularly as epitomized by learning regions or cities (Maskell and Malmberg 1999). More recently, the focus has shifted to the diverse means of knowledge transfer, whether localized or 'distanciated' (Amin and Cohendet 2004).

The notion of distanciated knowledge implicitly recognizes the role of exogenously sourced knowledge, and this is linked to the long-established literature on the role of foreign direct investment and trade as sources of knowledge, competitive advantage and growth. The classic model of the use and re-use of knowledge within transnational companies dates from the early writings on industrial organization in multinational enterprises (Hymer 1960; Kindleberger 1969). These argue that foreign investors enjoy absolute, ownership-specific advantage over firms in the host country – which both drives and facilitates foreign direct investment. Trade is also classically used as a surrogate measure of knowledge transfer in many economic analyses – buying goods is, in a way, buying technology and other forms of knowledge, whether directly as technology and capital goods, or indirectly in the form of other products.

This literature on multinationals is of less importance to this volume than the understanding that person to person contacts are also potentially important channels of knowledge transfer. Of course, such expertise or knowledge can be transferred, at least in part, electronically, whether by email, interactive web sites, or videoconferencing. But, as will be argued in Chapters 2 and 3, human mobility is a highly effective, and distinctive, channel for knowledge transfer. Moreover, for some forms of tacit knowledge, it may be the only means of transfer. This brings us to the central concern of this volume, which is the role of international mobility and migration in the transfer and connection of knowledge, and learning. This is still surprisingly under-researched, although – as will be argued in Chapters 2 and 3 in particular – there has long been an implicit, if not an explicit recognition of this important relationship. It is a relationship which has moved to the top of many policy agendas, epitomized by concerns about the so-called 'global talent wars'.

Global talent wars

The role of migrants in knowledge creation and transfer has to be seen in context of globalization and demographic ageing tendencies in the more developed countries. This leads to intensified competition over attracting pools of potentially mobile knowledgeable migrants, which has rather colourfully but effectively been described in terms of 'global talent wars'. The OECD (2002) among others, has recognized that there is, or potentially will be, a global talent war, as national governments compete against each other to secure the skills and knowledge of migrants. The same applies to particular regional and urban economies, and to firms, that increasingly are involved in the global sourcing of exogenous talent through international migration. This applies not only to the major western economies, but also to the dynamic economies of South-East and Eastern Asia. For example, commenting on the position of Singapore, Fong (2006: 155) quotes from a speech made by the then Prime Minister, Goh Chok Tong, at a National Day Rally in 1997, emphasizing that foreign talent is:

> a matter of life and death for us in the long term. . . . If we do not top up our talent pool from the outside, in ten years' time, many of the high-valued jobs we do now will migrate to China and elsewhere, for lack of sufficient talent here.

It is the sense of urgency, and the implication of fundamental economic consequences, which has fuelled and intensified the so-called talent wars which Clegg (2007: 10), perhaps over-dramatically, extends into a powerful metaphor: 'Often likened to a military struggle, the war for talent gives every sign of spiralling into an arms race' (Clegg 2007: 10).

In a theme that recurs throughout the book, we argue that it is not only the obviously highly-skilled workers, such as doctors, IT experts and scientists who are in global demand. Instead, there is a wide, and growing, range of occupations which potentially are the objects of global talent wars, including football players, opera singers, and chefs (see Box 1.1). At a less intense level of demand, there is also strong competition for those with 'intermediate' skills, such as electricians, conservation workers, or nurses. Moreover, given that skills are socially defined, many so-called 'unskilled jobs' actually have relatively high knowledge contents. In addition, many highly-skilled migrants for a variety of reasons take jobs abroad – at least initially – which are relatively unskilled, so that they represent potentially substantial knowledge reservoirs in the destination country, or in the country of origin if they can be enticed to return. These arguments serve to underline the actual and potential contributions of international migration to knowledge in a range of sectors, and territorial economies – particularly in removing or ameliorating constraints in the supply of skills and knowledge.

While our emphasis on international migration stems from recognition of expertise (that is possession and application of knowledge) as a key component of economic development, this has been reinforced by structural shifts in the economy, especially the expansion of the service sector and increasing international trade in some types of services, notably business services. As Millar and Salt (2007: 43) argue: 'The main factor of production involved in service trade is expert knowledge: trade in services is trade in expertise'. This opens up several possibilities, including – most commonly – the export of services produced by endogenous labour. Our concern, however, is with how these structural shifts in the economy are leading to new demands for the global sourcing of expertise, talent and knowledge, whether by firms or by territories.

Given that the national state is still the most significant site of regulation of international migration, national 'migration policies' (understood

---

Box 1.1 **The international culinary talent wars**

Against a background of a shortage of highly skilled chefs in the UK – some 40 per cent do not possess a Level Two qualification which is deemed to be the minimum needed to prepare food from raw materials – the country is becoming engaged in an international culinary talent war. Bob Cotton, the British Hospitality Association's chief executive said at a launch event: 'With our food tastes evolving, there is an urgent need for great chefs, and this will become even more of an issue with the influx of millions of tourists heading to London in advance of the Olympic Games'.

Cotton proceeded to argue that:

> Foreign chefs wanting to work in Britain should be given the same fast-track treatment as top footballers. . . . Highly experienced chefs from India, China, Japan and elsewhere were 'queuing up' to move to the UK. . . . There is a strong market for chefs who are based in the EU and move from place to place. However, the high quality Asian restaurants can't find enough locally and need to look abroad.

Brian Wisdom, chief executive of People 1st, echoed these comments, saying:

> The government has known for some time that employers here are struggling to find highly skilled chefs. . . . There is a certain irony in the fact that a sushi chef with 12 years' training – who we really need in this country – gets denied entry, yet footballers from the same part of the world with less years' training behind them take priority.

Source: after *Financial Times* (2007); also available online on: http://www.personneltoday.com (accessed 28 November 2007)

---

to include a range of linked employment and welfare policies, as well as narrow immigration laws) have become a major battleground in the global talent wars. They are a battleground in two senses. First, as the principal vehicles through which states intervene in the global talent wars, in support of national or domestic capital, and national growth strategies. Second, because migration policies are not shaped only by the dictates of national and international capital, but by the outcome of a political process, which engage competing interests ranging from endogenous workers seeking labour market protection from international migrants, to political parties responding to a range of electoral concerns and prejudices. The outcome is a bifurcation of national migration policies, between those that engage positively with 'skilled workers' as opposed to the generally restrictive approach to other migrants. As Findlay (2006: 68) writes:

> it is interesting to note the curious contradictions of contemporary migration policy statements. At a time when European governments seem to be competing with each other to emphasize how firmly their borders are closed to so-called 'bogus' asylum seekers, there also appears to be a new scramble to recruit highly skilled workers. Thus, while the electorate in most West European democracies are wooed with messages about the front door of the state being firmly bolted to clandestine migrants and unwanted entrants. . . . Governments have performed a remarkable change of policy over the last ten years with regard to highly skilled migration. . . . Promoting skilled immigration it has been argued is not only highly appropriate because of low levels of demographic growth, but more significantly it is a prerequisite to sustained economic growth in a competitive global economy.

Global talent wars are usually considered as being the domain of skilled or highly-skilled workers. Unskilled workers are, at best, tolerated as a social and economic necessity to fill particular jobs, often the dirty or low-paid jobs that are required in major cities to support the elite workforce (including highly-skilled migrants). Sassen (2000a) powerfully stressed this argument in her discourse on the dual streams of migration that sustain the economies of global cities. The polarization of migration into two types is, of course, an analytical device, and it does not capture the complex relationships between migration and knowledge. Migrants are no more divisible into two types – skilled and unskilled – than the economy is bifurcated into the knowledge-based economy and the non knowledge-based economy. But the energetic revamping of national migration policies in recent years, as part of the talent wars, is a powerful expression of the way that these have moved up the policy agenda in many, if not most of the more developed countries.

What's new? The historical context

While emphasizing the important role of international migration in knowledge creation and transfer, this book does not wish to exaggerate its significance. Knowledge is not the only driver of economic growth and competitiveness, and international migration is not the only, or even the most important, channel of knowledge transfer in many situations. We also acknowledge the critique of the 'newness' of large-scale international migration, and of the significance of migration-transferred knowledge.

First, consider the critique of the novelty of large-scale migration – with some 200 million people now being estimated to be working outside their countries of birth. The question is whether this constitutes an unprecedented level of international migration. The answer to this disarmingly simple question is inevitably complex. There are considerable difficulties inherent in the historical analysis of international migration, and reliable data over more than a few decades are only available for a few countries (but see Box 1.2). However, even putting aside some of the formative migration flows in the early history of population settlement, a historical analysis emphasizes that, in many ways, the high levels of international migration in recent decades are not as unprecedented as is sometimes implied. It is particularly instructive to compare the late twentieth century to the second half of the nineteenth century and the early twentieth century, an era of significant international migration. Chiswick and Hatton (2003) provide a useful comparison of migration in these two time periods. They estimate that in the three decades after 1846 there were about 300,000 migrants per annum from Europe – the source for which emigration was probably best documented. This was a period when the earlier dominant migration stream from the British Isles was joined by new streams of migrants from Germany and, after 1870, from Scandinavia and elsewhere in north-western Europe. After the 1880s there was a new surge of emigration, from Italy and the disintegrating Austro-Hungarian empire, followed by southern and eastern Europe. Gross emigration numbers climbed to circa 700,000 a year by the late nineteenth century, and peaked in the first decade of the twentieth century at almost one-and-a-half million a year. In addition, there were also major migration movements within Europe, such as those from Ireland to Great Britain, and from Asia to East Africa, the Pacific Islands and the western regions of North America.

After the First World War, a combination of economic depression and the introduction of immigration quotas in the USA sharply reduced international migration. This was followed after the Second World War by the period that Chiswick and Hatton characterize as 'Constrained mass migration, 1946–2000'. In this period, gross immigration to the USA, Canada, Australia and New Zealand alone climbed to over one million per annum by the 1990s. At the same time, there was a shift in the source of global migrants from Europe to Asia, Latin America, Africa and the Middle East. While absolute numbers increased compared to the late nineteenth and early twentieth centuries, population numbers also increased so that, in overall relative terms, international migration was probably no greater in the late twentieth than the late nineteenth century.

Although the relative numbers may not have changed significantly, there have been other changes in the nature of international migration. These shifts are related in part to the growth of refugee movements, but also to changes in the origins of migrants, which have often led to the racialization of the politics of immigration. Castles and Miller (2003: 1) summarize these: 'while movements of people across borders have shaped states and societies since time immemorial, what is distinctive in recent years is their global scope, their centrality to domestic and international politics and their enormous economic and social consequences' (2003: 1).

The globalization of migration has several different sources (Abella 2006: 11), including the following. First, the growth of transnational companies with global supply and distribution chains, requiring inter-company mobility

Box 1.2 Migration to the Americas in the 15th–19th centuries

The discovery of the Americas by Columbus started one of the greatest migration flows in human history. The New World experienced severe labour shortages, as a result of territorial and economic expansions driven by the colonial powers. These colonial powers responded to these shortages through massive transfers of slave labour, above all, but also of convicts and indentured servants. Some 11.3 million people arrived in the Americas from Europe and Africa in the period 1492–1820. The three above-mentioned groups accounted for 82.3 per cent of the total migrants, with forced slave migration being particularly important. The composition of the labour inflows only changed after the abolition of slavery. In the period 1820–80 some 16 million people arrived in the Americas, of whom some 81 per cent were 'free' migrants, even though this often implied only the freedom to take decisions in the face of acute economic, religious or political pressures in their countries of origin.

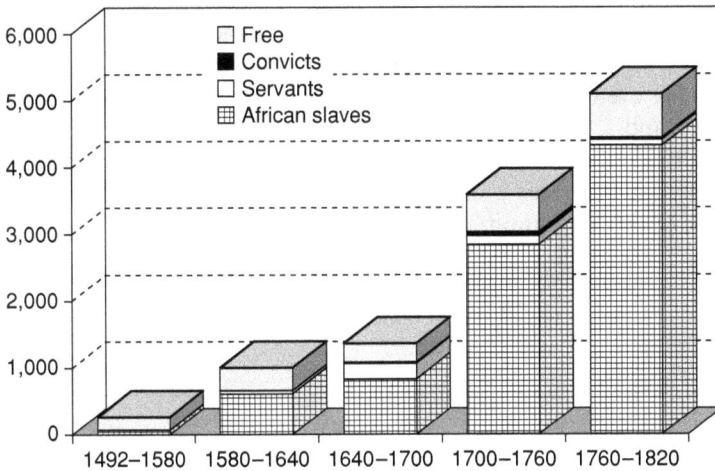

Figure 1.1 Classification of arrivals in the Americas, 1492–1820 ('000s)

Source: after Hatton and Williamson (2006)

of managers, professional and technical workers. Second, the demand for migrant workers willing to undertake those dirty and low-paid jobs that are increasingly shunned by domestic workers. Third, the seemingly explosive growth in demand for skilled workers in particular sectors, especially IT but also biotechnology and health. Fourth, the growing numbers of students studying abroad, who decide not to return to their countries of origin thereby swelling the numbers of international migrants; this has been facilitated by favourable visa regimes in several countries, designed to encourage staying on. Fifth, large numbers have been displaced by political and military conflicts and instability, drought, famine and dire poverty. And finally, there are demands for increased numbers of health and care workers, as the populations of the developed world age, and they spend increasing proportions of public and private funds on such services.

There have also been shifts in the skills composition of the migrants (Chiswick and Hatton 2003: 93–4). Abella (2006: 13) addresses the question of whether the international mobility of the highly skilled has risen significantly in recent decades and concludes that, although deficiencies in the available statistical data on migration make this problematic, 'most observers seem agreed that this is the case'. Docquier and Rapoport (2004) also provide statistical evidence of the higher propensities of the skilled and highly skilled to migrate in recent years, although their evidence is limited to a few more developed economies (Box 1.3).

Box 1.3 Immigration is becoming more skilled

More skilled workers are more likely than less skilled workers to become migrants. Estimates of average global emigration suggest that the rates are 0.94 per cent for low-skilled, 1.64 per cent for medium-skilled and 5.47 per cent for highly-skilled workers. In other words, there is a ratio of more than five to one in their propensity to migrate internationally. This can be explained in terms of both differential resources for migration, and the selectiveness of most national immigration policies.

Table 1.1 Admissions of skilled immigrants, 1991 and 2001

Country	Number ('000s)	
	1991	*2001*
Australia	41	54
Canada	41	137
United States	12	175
Sweden	0	4
United Kingdom	4	40

These data clearly demonstrate a sharp increase in skilled immigration in the 1990s. In addition, there has also been the growth of skilled immigration via temporary schemes, notably in the USA where such numbers increased from 143,000 in 1992 to just over 500,000 in 2000.

Source: after Docquier and Rapoport (2004)

The consequences of these migration changes are highly unevenly distributed. Although skilled migration flows are on the increase, the largest flows of highly-skilled migrants are highly concentrated among a small number of more developed economies, especially within the EU countries and between the EU and North America (Abella 2006: 13). However, the greatest consequences are felt not in these countries – where migrants still represent relatively small proportions of the skilled labour force – but in less developed countries, where the skilled emigrants usually represent substantial proportions of their total reserves of skilled workers (see the discussion of 'brain redistribution' in Chapter 2).

The outcome of these changes is that, as Kuptsch and Fong (2006: 8) argue:

> Global talent has never been more mobile or sought after. A complex phenomenon that takes many forms, it comprises many groups of people – temporary skilled migrants moving to take jobs of limited duration, refugees, skilled immigrants, students and even tourists – whose movement across borders augments a receiving country's stock of human and technological skills. Opportunities for such workers have expanded with globalization, and barriers to their cross-border movement have fallen as countries actively promote inflows to redress domestic skill shortages and to quicken economic growth.

Therefore, the answer to the question of 'what's new' lies as much in the diversity and complexity of the flows of people across borders and into economies, as in the numbers involved. There has also been a marked shift in demand and policies towards the more skilled spectrum of the workforce. However, the answer to 'what's new' also partly lies not only in the scale and nature of migration, but also in the nature of knowledge, and the increasing emphasis placed on tacit knowledge in the economy. This provides the context within which the economic consequences of international migration are played out, as well as shaping the scale and selectiveness of international migration, and migration policies.

The increased importance of knowledge in modern economies is rooted in a number of analyses of 'postindustrial', 'post-modern', and 'knowledge' societies (reviewed in Drucker 1993) (see also Chapter 3). There are several distinct strands of the theme of the growing importance of knowledge. First, there is an increasing emphasis on knowledge in determining the capacity of firms to respond rapidly in increasingly competitive conditions, exacerbated by globalization. Second, new technologies provide means to generate, store and transfer information at a scale and speed that would have been unimaginable even two decades earlier. Although such information is codified, and easily transferable electronically, it requires the application of tacit knowledge possessed by individuals (and also, arguably, the collective knowledge held by organizations) in order to utilize information effectively. Third, there has been the relative growth of knowledge-intensive sectors, particularly in the service sector: although Drucker was not specific in his definition of these sectors, some international definitions, such as that used by the OECD, define the knowledge-based economy as those sectors where 25 per cent of the labour force are graduates. Drawing these three strands together, Drucker (1993: 38) reached his much-quoted conclusion that 'knowledge is the only meaningful resource today'.

Of course, knowledge has to be combined with capital and labour, but is also indivisible from these, having shaped the flows and stocks of both. Knowledge has always been integral to all forms of economic activity and it is impossible to conceive of economic activity at any point in history when knowledge was not crucial to its creation, operation and expansion. However, there have been qualitative changes in the volume of information and knowledge available, and in the speed of dissemination. And, despite the emphasis on the collation, transfer and transformation of electronic databases, tacit knowledge or personal knowledge (Polanyi 1958, 1966) remains of enduring, and arguably of enhanced, importance in the modern economy (Nonaka *et al.* 2001: 490).

The arguments about the changing role of knowledge in the economy have been given an additional twist since Drucker's work by the concept of the knowledge-based economy. As Jessop (2006) argues, this is a performative paradigm, meaning that it brings into existence what it purports to describe. In particular, it is a powerful economic imaginary, behind which powerful vested interests have marshalled, to bring about a policy and intellectual paradigm shift which have emphasized the role of selective knowledge-based sectors as the drivers of economic growth and development. There are several reasons for this paradigm shift including the fact that the new economy often does not seem to have a material base, its link to the powerful imaginary of the information economy (Castells 1996), greater awareness of intensified international competition, and the search by international organizations for new policy frameworks.

Of course, the knowledge-based economy no more represents the majority of jobs in any economy, than it is the exclusive preserve of such jobs. Rather, highly-skilled workers are to be found in all economic sectors. Although technological changes may lead to deskilling in some sectors (for example, the impact of chill-cook and microwave ovens in restaurants), there are usually also examples of new technologies and new processes of production requiring increased skills in all sectors. The stereotype of the highly-skilled migrant may be the financial analyst or the brain surgeon, but is also exactly that, a stereotype which tells only part of the story of international migration and knowledge in the modern economy. As will be argued later in this chapter, all workers are knowledgeable. This is as true of the au pair as it is of the IT specialist, and of the plumber as it is of the brain surgeon. Although highly-skilled workers are distinguished by the knowledge they have accumulated, all workers possess some tacit knowledge.

From the skills dichotomy to the knowledgeable migrant: the approach of this volume

Having described the background to this book, and some of the key issues that frame the relationship between migration and knowledge, this final section of the chapter sets out the broad ambitions of the volume. The central focus of the book is the role of international migration or mobility in the transfer and application of knowledge in the economy. More precisely, we assert that if knowledge is the key driver of economic change, and much of that knowledge is transferable, then migration is an important, and in some respects unique, channel for knowledge transfer and application. Although there is a considerable body of research on, on the one hand, human capital and skills and, on the other hand, mobility and migration, there are still surprisingly few studies of the role of human mobility (see Argote and Ingram 2000) – let alone international migration – in creating and disseminating knowledge in the economy.

Three features of our approach to the relationship between migration and knowledge require further clarification at this stage: our understanding of international migration; the socially situated nature of knowledge and knowledge sharing; and – most importantly – our advocacy of a holistic and unified approach which places the knowledgeable migrant at the centre of our analysis, rather than the arbitrary and polarized notions of skilled and unskilled migrants.

First, we need to consider *what is understood by 'international migration'*. This assumes many forms – including the migration of non-economically active dependents, or later life migration – but here we focus on international mobility which is economically motivated and/or results in economic activity. Therefore, we are interested in refugees but only if they are economically active – which they may not be because of conditions attached to their migrant status.

There are also other issues relating to the definition of 'international migrants'. International migration has traditionally been considered to involve crossing an international border and a degree of permanence (to distinguish it from tourism, for example) (Boyle *et al.* 1998: 34). There is no theoretically grounded definition of 'permanence', so that a necessarily arbitrary minimum period (usually one year) has been used in most secondary data sources. But that does not take into account the emergence of new forms of migration, notably circulation and temporary migration (King 2002), with migrants working abroad for shorter and more discontinuous time periods (see Chapter 4). Attempting to place time limits on different types of migration, or on migration as opposed to visits, is a futile exercise, as this is more a matter of meanings and material consequences (for example, opening local bank accounts, or housing arrangements) than of the numbers of days spent in another country. Instead, temporary migration can be understood to exclude business trips and visits, but include short-term work assignments where these involve significant disruption to previous or normal working and living arrangements, even though these are not permanent. Permanent migration involves a more prolonged period working in another country, probably involving more durable and significant changes in living arrangements and material circumstances in both the country of origin and destination. These definitions are starting rather than end points for our approach in this book because the fluidity of mobility types, and shifts in individual mobility experiences through the course of the life cycle, defy easy generalization. Moreover, there is a dynamic relationship between types of mobility and the changing role of different types of knowledge used in the economy – they are mutually informing. New forms of mobility partly evolve to meet the changing knowledge requirements of firms, but the emergence of new forms of mobility can also inform learning and the way in which knowledge is transferred in these economies.

The book is also concerned with the full cycle of migration and with transnationalism. Traditionally, migration studies tended to view migration as a relatively simple single movement, usually resulting in long-term or permanent migration and 'settlement'. This led to a relative neglect of return migration (but see King 1986a), which became unsustainable in the face of shifts to more short-term and cyclical mobility (King 2002). Migrants increasingly undertook temporary or short-term migrations involving returns to their countries of origin, well before the end of their economically active lives. The picture became even more complex as migrants became involved in serial migrations (Ossman 2004), or multiple cycles of migration rather than one cycle of migration. The economic consequences of migration and mobility are played out over the life course, with the experiences of migration and return migration being interwoven with other aspects of the individuals' lives.

Furthermore, although the book is entitled 'international migration', it also engages with those working/living practices and identity formation and reformulation processes that constitute transnationalism (Vertovec 2004). Transnational migrants are actively engaged in networking within and between more than one place, and their identities are often hybrid, as they work and rework their identities around different places. Zhou and Tseng concisely express this: they 'build fields that link together countries of origins and destination' (Zhou and Tseng 2002: 132). This is not to say that all migrants lead transnational lives, but this does represent one important strand in the experiences of many migrants.

Second, following Blackler (2002: 63) the book is infused by the notion that 'knowing should be studied as practice, and practice should be studied as activity that is rooted in time and culture'. In other words, *knowing, learning and knowledge are socially situated* (see also Brown and Duguid 1991). The knowledge that migrants acquire, share and use is shaped by their relationships with significant others. At one level, these significant others include most obviously the workplace and relationships with other migrants, indigenous workers, and managers. But the boundaries of modern workplaces are blurred and significant relationships may also be formed with individuals in other organizations. Similarly, individuals may be embedded in formal or informal associations and networks as communities of practice (Wenger 1998). These may be highly localized, or within single firms, but they can also extend across international boundaries, including the places where migrants formerly lived and worked. And those who lead genuinely transnational lives may be engaged in multiple knowledge transactions with other individuals widely distributed between multiple places.

Socially situated knowledge is also shaped by the multiple levels within which individual lives are played out. These range from the family to the local community, to the urban or regional, the national and the global. Specific institutions and practices within each of these levels shape knowledge and migration. While all these levels are important in different ways, this book focuses mainly on four key sites: the individual (Chapter 7), the firm (Chapter 6), the region and the national state (Chapter 5).

- There is a powerful argument for front-staging the individual in analyses of migration and knowledge (Folbre 2001; Williams 2006a). It is the individual migrant rather than the family or group of migrants who is the fulcrum of knowledge transactions in the modern economy – although this may have been less true in earlier periods when the family was the unit of production, as in the migration to the Americas before the mid-nineteenth century. The knowledge held by individual migrants, and their ability to share and utilize this, is determined by their lifetime experiences as well as by a range of socially-mediated individual dispositions such as risk-taking propensity and intelligence.
- The focus on the firm (understood as production units of different types, whether in the private, public, social or voluntary sectors) follows from our central interest in economic activities. This is where individual workers engage to different degrees with co-workers and managers. Knowledge management is, in fact, a fertile area of research in economics and business studies, although there has been surprisingly little research on migrants in this context, excepting ethnic businesses and self-employment. It is, above all, a site of knowledge sharing.
- The regional level is significant because many of the institutions that shape learning in firms are articulated at this level – whether in terms of the activities of local and regional authorities in the fields of education and training, or the fostering of high levels of trust and collective learning. Regions are also the level where migration intersects with the dynamics of local labour markets, and where the impacts of international migration on endogenous labour are played out. The literatures on clusters and learning regions also point to the region as a key site for inter-firm knowledge sharing and spillovers. Additionally, there are also arguments that some regions, and particularly metropolitan areas, offer particularly creative cosmopolitan social milieu (Florida 2005): these are shaped by

migration, attract migrants and mediate their knowledge creation and transfers. Hitherto, undue attention has been given to the role of migrants in particular regions, such as Silicon Valley or London, as iconic representations of the symbiotic links between creativity and highly-skilled migration. But this is only part of the story, of course. The regional level is variably constituted and international labour migration plays a role in a range of regions, both metropolitan and rural, as well as the dynamic and the laggard.

- Finally, despite globalization tendencies, the national remains a key site for regulating economic activity (Hirst and Thompson 1992), for example in terms of the national innovation strategy, as well as migration, particularly in respect of immigration and employment, but also the national innovation system. The experiences of national territories differ dramatically not only between more and less developed countries, but even within supra-national economic blocks such as the European Union.

Although the book considers each of these levels separately in Chapters 5, 6 and 7, this is for analytical convenience. In practice, they are strongly inter-related, and we concur with Amin's (2002: 386) view that levels are co-constituted and folded together. The three levels are inter-related in complex ways. For example, the national level determines labour market entry conditions for migrants but, in turn, national regulation is shaped by the requirements of the corporate sector, or by the broader dictates of capital accumulation in particular regions (for example, the labour needs of London in relation to UK immigration policies). While migration intersects with local labour-market dynamics at the regional level, these dynamics are constituted of individual firm behaviour and are shaped by national regulation. Finally, workplaces are critical domains wherein migrant knowledge is commodified, and tacit knowledge is shared among employees. But national employment and taxation laws shape firms' reliance on internal versus external labour-market strategies, and hence international migrants.

Third, our approach is to see *all migrants as knowledge bearers* with the potential if not always the experience of knowledge sharing and learning. Similarly, we consider that migrants play a role in knowledge transfer in all or most economic sectors. Of course, their potential and their contribution is highly unequal between individuals, occupations and industrial groups, regions and states. But such a holistic approach is necessary to counteract the elitist approach of existing research, which has tended to focus on elite workers, in elite industries in elite places (Williams 2006c). The literature on skills and knowledge largely focuses on the so-called highly-skilled migrants, owners of intellectual capital who have potential to add to creativity and productivity, particularly in the 'knowledge-based economy', and its iconic territorial articulation in places such as Silicon Valley or Cambridge at the regional levels. In contrast, much of the literature about the economic impact of 'unskilled migrants' portrays them as empty knowledge vessels who, depending on the politicization of the argument, either threaten the wages of other unskilled workers or provide the essential staffing of the services that support knowledge economy workers in cities such as London and New York (Sassen 2000a). Unskilled migrants – a term that itself denies their particular skills – are seen as replacement labour, as constituting part of low-wage, occupationally-segmented labour markets (Piore 1979) or as a source of 'just in time' workers (Bryant *et al.* 2006). In contrast, we argue that there is a continuum of skills and knowledge, and that over time individual migrants can be relocated within this.

This polarization in the research literature resonates with Robinson and Carey's (2000: 103) view that the prevalent dichotomy between unskilled and skilled is 'artificial and unhelpful, giving undue salience to a single characteristic of the individual'. Moreover, there have been deep changes in the organization of work, so that so-called 'unskilled' jobs increasingly require a range of 'social skills', such as communication, or team working (Payne 2000). And Florida (2005: 4–5), writing about creative cities and regions, also contends 'that every human being is creative'. This can be paraphrased to argue that every migrant is a learner, knowledge carrier and knowledge creator: the extent of this, and its economic impact, may vary considerably, but the underlying processes of learning and knowledge transfer are shared to some extent. Care must be taken, however, not to glamorize jobs that may involve drudgery, and routine tasks, even if they also involve some learning and knowledge transfer experiences.

This perspective links to critiques of the highly gendered nature of research, and policy debates, on 'skilled' migration. As Kofman and Raghuram (2005: 150–1) argue, the neglect of women in the literature on skilled migration partly arises from the problematic definition of skills. The emphasis on technological innovation, and 'the new knowledge economy', has focused attention on scientific and technological jobs, thereby ignoring the skills required in – for example – educational and caring jobs, such as teaching and nursing. In reality, as Williams and Baláž (2004) demonstrate in their study of Slovak au pairs in the UK, there is a vast amount of learning and knowledge creation not only in the public sphere, but also within the private sphere of the home, which potentially can be commodified in the labour market (for example, social or language skills).

While the above stereotypes are exaggerations, they do illustrate a real divide in the approach of most research to skilled vs unskilled migrant labour. There are of course significant skills and knowledge differentials between these two polar types of migrants, but the bifurcated nature of the literature – empirical studies rarely consider the two groups together – means that a number of essential similarities are overlooked in respect of both the migrants themselves and the environments they work in. In a sense, this replicates similar fault lines, which have appeared in the generic research on knowledge and knowledgeable workers. For example, as Khadria (1999: 25) reports, it was Peter Drucker who 'introduced a classification of work among two divergent employment categories, viz, knowledge workers and service workers'.

Thompson *et al.* (2001) critique the concept of the 'knowledge worker' with its selective association with the knowledge-based economy, and denial of the importance of knowledge to the work of all workers. They propose as an alternative the notion of 'the knowledgeable worker'. By extension the focus of this book is what we have termed 'the knowledgeable migrant'.

Conclusions: the socially situated knowledgeable migrant

As emphasized in the previous section, the central concern of the book is the role of international migration or mobility in the transfer and application of knowledge in the economy. Above all this is based on our argument – elaborated at length in Chapter 3 – that tacit knowledge is increasingly important in the modern economy, and that migration is an important and sometimes the only channel for the transfer of such knowledge. Much of the literature on migration and skills has tended to analyze skilled and unskilled migrants as different categories, requiring different theoretical lenses, and posing different policy and conceptualization issues.

In contrast, our approach is to see all migrants as knowledge bearers with the potential, not always fulfilled, for knowledge sharing and learning. This is also linked to our argument to avoid undue focus on highly-skilled workers in iconic sectors, working in iconic places, such as the IT sector in Silicon Valley. Of course, skilled and unskilled workers have differentiated economic impacts, and an IT expert working in California has far greater potential to be creative, and to transfer distinctive knowledge, than does say the migrant working on an assembly line in a rustbelt economy. Nevertheless, adopting Thompson *et al.*'s (2001) concept of the 'knowledgeable worker', all migrants can be seen as 'knowledgeable migrants'. Moreover, we also focus on the full migration cycle, and on the socially situated nature of knowledge and learning.

2 Theorizing international migration and knowledge

Theoretical perspectives: from human capital to knowledge(s)

There are several different conceptualizations of skills and knowledge, and these have been adopted in the literature on international migration in a broadly sequential manner (Figure 2.1). The starting point for theorization is human capital theories, which originated in neo-classical economics. These theories draw on a strongly deductive theoretical framework, but have been explored and challenged by extensive empirical econometric research. The theories are essentially based on individual decision making under assumed conditions, and they initially focused largely on 'permanent' labour migration. However, more recent research has focused more on questions of return and temporary migration, recognizing some of the significant shifts in the nature of international migration (see Chapter 4). There has also been a sustained debate over the aggregate effects of international migration on human capital stocks in destination and origin countries, a debate that started with the notion of 'brain gain' and 'brain drain', but which has subsequently been broadened to consider the complexities of these human capital 'redistribution' or 'brain distribution' effects (Williams and Baláž 2005a).

There has been a growing critique of the limitations of human capital theory, encapsulated in the notion that there is a need for a broader understanding of the full range of skills possessed and acquired by migrants, that is the need to take into account 'total human capital' (Li et al. 1996). That has led to an argument that researchers should focus not so much on skills, with the attendant limitations of how these are assessed in terms of qualifications, education, or technical skills. Rather, there is a need to look at the concept of 'competences', which directs attention to a range of so-called soft skills, such as interpersonal abilities, confidence, or communication skills. This is more than just a debate about terminology, because a focus on competences can lead to different assessments of the contributions of migrant workers, and in particular of their learning experiences (Williams and Baláž 2005a).

More recently, one of the authors has argued for a need to focus not so much on the skills of individual migrants, as on their knowledge (Williams 2005; 2006a; 2007a; 2007b). In some ways this builds on the notion of competences because definitions of different types of knowledge (Blackler et al.

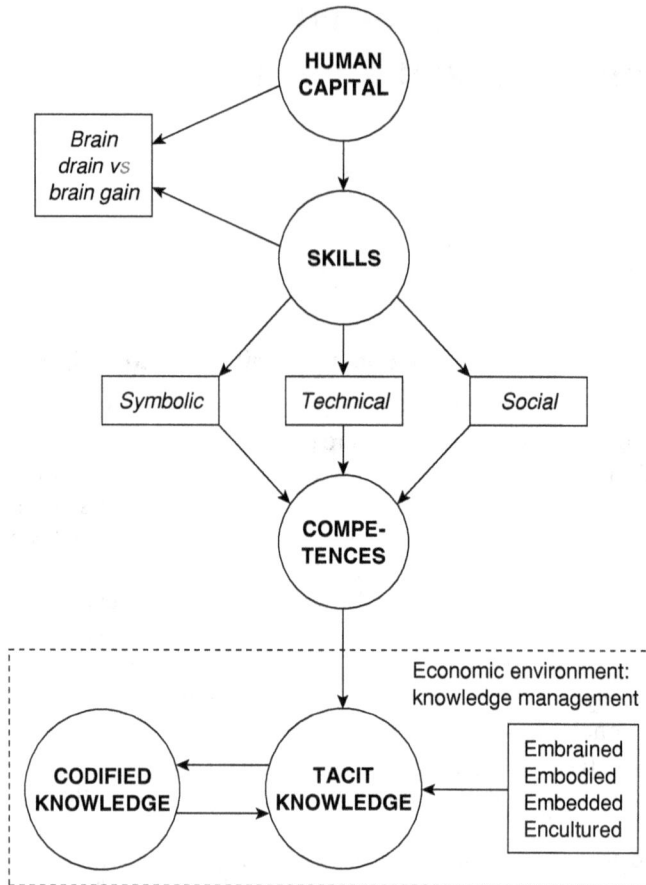

Figure 2.1 Theoretical perspectives: human capital, competences and knowledge
Source: authors

1998) re-emphasize the need to escape from narrow, technically-based or qualifications-based assessments of skills. However, the focus on knowledge – which exists in both explicit and tacit forms, at individual and organizational levels, and within and without organizations – allows migrants' roles to be placed in fuller perspective of the total system of knowledge in an economy.

Finally, we draw on one other, and very different strand of research, labour market segmentation, as a way of examining the socially situated nature of migrants' experiences in the workplace. That is, how particular labour market practices shape their learning and knowledge transfer experiences. This provides a theoretical background against which we explore more specific issues in the workplace in Chapter 7.

There are, of course, other theories of migration, particularly those that focus on behavioural complexities, networks, and identities (Brettel 2000). These do not necessarily contradict some of the ideas explored here, rather they provide broader and often complementary perspectives on understanding migrant decision making and behaviour. But all these theories share a common focus on the selectiveness of emigration, for example that migrants are more likely to be risk taking or adventurous, which in turn may have implications for learning and knowledge transfer. We will draw, therefore, on some of these theoretical insights later in the volume, but here we focus more narrowly on human capital theories and skills because of their particular theoretical proximity to the issues of knowledge transfer that we explore in this book.

Human capital theories and skills: returns, costs and individual decision making

The essence of these theories is that human capital is the main driver of economic growth, and this was crystallized by writers such as Becker (1964), although the concept dates back to the work of Adam Smith. Human capital theorists consider expenditure on education and training to be investments in *human* capital, so called because – unlike say personal financial assets – individuals cannot be separated from their knowledge or skills. It is a theoretical perspective that has shaped policy debates, as well as being driven by these concerns. As Lloyd and Payne (2003: 85) comment: 'Since the late 1980s, a global policy discourse surrounding education and training has been in evidence across the advanced industrial world, albeit one that plays out rather differently across countries'. Human capital is increasingly seen as a, if not the, key to economic growth and competitiveness, and national states have sought out various strategies to enhance what has come to be seen as a vital resource.

In the simplest terms, human capital stocks can be increased in one of two main ways: through education and training (that is through measures directed at indigenous, or potential indigenous labour), or through immigration. Perhaps this is rather too much of a simplification, because the available supply of human capital at any one time is also mediated by social institutions, including retirement and minimum working age rules, and deeply embedded working practices. Here we focus on migration, where two main questions have been addressed by human capital theories: the decision to migrate (and return), and the impact of migration on individual and aggregate welfare. Aggregate welfare we consider later in this chapter.

First, we consider what human capital theories say about individual decision making. Human capital theories assume that there is a relationship between the individual's investment in education and returns to this, mainly in terms of wages although socio-psychological returns are also recognized. Initially, the migration decision was understood, within this framework, as an investment of resources in moving to a country where higher wages are available, but of course – as with most investments – there are known and unknown costs and risks. Therefore, later human capital models conceptualized migration as an investment decision whereby individuals calculate the expected future discounted returns (benefits minus costs, discounted to present values) of migrating or staying put, taking into account risk as a component of costs (Stark 1991). Sjaastad (1962) provided a perspective on the costs of migration, in context of migration being considered a resource decision (Box 2.1). The propensity to migrate also varies according to the income available, whether from wages, land or owning a business as these determine the net return to migration (Massey *et al*. 1993: 456).

Box 2.1 **The costs of migration in context of a resource decision**

In one of the seminal contributions in this field, Sjaastad argued that migration can be seen as a resource allocation problem. It is 'an investment increasing the productivity of human resources, an investment which has costs and which also renders returns' (p. 82). There are both money and non-money costs: while the former include the expenses incurred in moving, the latter include (unknown) earnings that have been forgone, including the opportunity costs of spending time looking for a new job. But Sjaastad also recognized that there were 'psychic' costs of moving to a different 'environment'. These are costs associated with leaving familiar surroundings, family, and friends. He even acknowledges that the non-money costs may be greater than the money costs. From a theoretical perspective he argues that:

> it would be difficult to quantify these costs; moreover, if they were quantified, they should be treated quite differently from the costs previously considered. The costs treated above represent real resource costs; however, the psychic costs do not. Rather they are of the nature of lost consumer (or producer) surplus on the part of the migrant (p. 85).

Source: after Sjaasatad (1962: 83)

This model has been widely employed to investigate migration decision making, initially in the USA but subsequently in a range of countries and contexts. Empirical research on human capital models has sought to explain the distribution of the migration/staying decision – who migrates and who stays – in relation to socio-demographic-economic characteristics. There is a long list of such characteristics, but they typically include age, education, occupation, employment status, language competence, sex, race, marital status and family size. Arguments are constructed as to why these should increase or decrease the rate of return or risk in the country of destination compared to the country of origin (Borjas 1987), and these relationships are explored using econometric methods.

One of the main issues to emerge from this literature is the question of whether migrants are positively or negatively selected – that is, do migrants with higher levels of human capital and other resources migrate because there is a greater premium paid to skilled workers abroad? Or is there a premium paid to such human capital in the home country, so that – especially when the costs of migration and preferences for consumption at home are taken into account – those with lower levels of human capital migrate? Massey *et al.* (1993: 456) consider this debate, starting with Taylor's (1987) work on Mexican migrants. This demonstrated that for rural Mexicans, the economic returns to investment in human capital (increased years of schooling) have been greater if they take jobs in Mexican cities than if they emigrate to the USA. This is because undocumented workers with a secondary education get the same low-wage jobs in California as do the migrants who have virtually no schooling at all. In other words, this provides evidence for negative selectivity.

However, as Massey *et al.* (1993: 456) argue, whether selectivity is negative or positive depends on the transferability of their human capital, 'which itself is determined by social, economic, and historical conditions specific to the countries involved'. Moreover, the market value of human capital in both the country of origin and the destination changes over time, so that it becomes 'nearly impossible' to predict the nature of selectivity as a function of the characteristics of individuals:

> In general, the only universal prediction that can be offered is that human capital should somehow be reliably related to the likelihood of international movement, but the strength and direction of the relationship is impossible to know in the absence of historical information about the countries involved.
>
> (Massey *et al.* 1993: 456)

This is necessarily a complex issue, and in Box 2.2 we have tried to summarize some of the available evidence.

***Box 2.2* Who moves? Human capital and migration selectivity**

If income differences were the only determinant of migration, it would be expected that migration flows would be dominated by low-skill, low-income migrants moving to high-income countries. However, an illiterate migrant has little to offer in a developed country and will find it difficult to obtain any but the most menial and low-paid of jobs. In practice, he or she would probably also find it difficult to obtain a visa to enter many countries.

Table 2.1 Migrants to the US by education level: expressed as percentages of the total educational cohorts still resident in these countries in 2000

	Education level			
	Total	*Primary or less*	*Secondary*	*Tertiary*
China	0.1	0.1	0.2	2.2
Indonesia	0.1	0.1	0.1	0.7
Philippines	3.6	0.6	2.2	11.7
Turkey	0.2	0.1	0.4	1.3
Brazil	0.2	0.1	0.5	1.1
Jamaica	33.3	4.7	40.9	367.6
Mexico	13.3	10.8	17.2	16.5
Peru	1.8	0.3	2.5	4.2
India	0.2	0.1	0.2	2.8
Pakistan	0.3	0.1	0.5	6.4

Source: after IOM (2005; Table 8.1); EC (2000); Meyer and Brown (1999)

Does this mean that migration favours the better educated? It is generally well-established in the migration literature that better-educated social groups tend to be the most internationally mobile (this may not apply to refugees). In the 1990s, the Developed Countries employed some 400,000 scientists and engineers from Developing Countries. This compared to 1.2 million scientists and engineers having similar jobs in developing countries (Meyer and Brown 1999). Jamaica was an extreme case: there were nearly four times as many Jamaicans with tertiary education in the US than in their country of birth (Table 2.1).

Migrants with secondary rather than tertiary education are favoured over those with primary or no education. Carrington and Detragiache (1998), for example, found that a majority of the migrants from Less Developed Countries have secondary education, and the migrants tended to be better educated than the population as a whole in the countries of origin. A study of push and pull factors in international migration in Turkey, Morocco, Egypt, Senegal and Ghana (EC 2000) also found that education tended to have a positive impact on the decision to migrate. In Turkey, Egypt and Ghana migrants have better education than non-migrants, even after allowing for age differences. In Senegal and Morocco, general education levels were so low that there was hardly any difference between migrants and non-migrants.

Sources: after IOM (2005, Table 8.1), EC (2000), Meyer and Brown (1999), Carrington and Detragiache (1998)

Second, turning to the returns to human capital from migration, there has been a focus on the wages paid to migrants compared to domestic workers in the destination economies. This is investigated by considering the wages of migrants compared to indigenous workers over time. Subsequent research has compared migrant earnings to those of locally-born ethnic minorities. The starting point in this research is that, following Becker, the human capital literature often distinguishes between 'specific' and 'generic' human capital. Specific human capital refers to skills or knowledge that is useful only to a single employer or industry, whereas general human capital (such as literacy) is useful to all employers. Economists view firms' specific human capital as inherently risky, since firm closure or industry decline lead to a surplus of skills that cannot be transferred (although it should be noted that the evidence on the quantitative importance of firm-specific capital is still unresolved).

Turning to migration, there is an assumption that migrant workers usually lack nationally-specific human capital, such as language skills, but also knowledge of local institutions and work practices. Critically, migrants incur costs related to a period of adjustment because, as Dustmann *et al.* (2003: 13) comment in their review of the UK evidence, the skills that migrants acquired in the country of origin may not necessarily or easily be transferable to the destination economy. This means that initially their productivity will be less than that of host-country workers who have broadly similar generic (i.e. excluding nationally specific knowledge) human capital, and their wages are therefore commensurately lower.

Over time, as migrants accumulate nationally specific human capital, their earnings increase, converging with those of indigenous workers, and this can be expressed in terms of an earnings function (Chiswick 1978; see Box 2.3). Much of the earlier research in this field assumed that migration was permanent: this fitted with the notion that the return to the migrants' human capital increased over time, and although the utility of the returns might decrease, his/her wage would still be higher than in the country of origin. More recent research has focused on return migration, a possibility that had not been engaged with by earlier theories. Dustmann and Weiss (2007), summarizing the work in this field, suggest three rationales for the decision to return within the framework of human capital theory, risks and returns. First, the balance between rent or income returns to human capital in the destination may change over time, while the migrants' preferences for consumption in the country of origin may also increase. Second, higher purchasing power in the country of origin, assuming either prices in general or specific prices such as housing, are lower in the countries of destination, means that their savings increase in real terms if they return. And, third, there may be a higher value placed on specific forms of human capital in the host country, especially in transition economies: for example, the migrants may have acquired particular technical or management skills, or language skills, for which a premium is paid in their countries of origin.

Box 2.3 **Human capital, length of residence and wages in the USA**

Chiswick's study of the relationship between migrants' human capital, length of residence and wages in the USA is one of the best known examples of research within the human capital tradition. Some of his key findings were that:

- In general, the annual earnings of foreign-born white men are lower than those of native-born workers, although by a small margin of only 1 per cent. However, holding constant some of the key indicators of human capital and employment (years of schooling, years of total labour-market experience, areas of residence and weeks worked), Chiswick found that the average weekly earnings of foreign-born workers are 3 per cent higher.
- Migrants' earnings are, however, significantly influenced by the length of residence in the USA. This is because over time they acquire an understanding of the language, culture and labour-market practices in the USA. This nationally specific human capital tends to increase their earnings over time; in other words, earnings can be expressed as a function of time since migration. Whereas immigrants on average initially earn less than native-born workers with similar socio-economic characteristics, their earnings rise sharply over time and the earnings gap between the two groups narrows. After five years, the migrants earn 10 per cent less than the native-born workers, after 13 years their earnings are approximately the same, and after 20 years they earn 6 per cent more. The 'earnings crossover' for most workers tends to occur after 10–15 years.
- The reasons for the crossover are complex but Chiswick suggests that they may 'be a consequence of a self-selection in migration in favour of high ability, highly motivated workers, and workers with low discount rates for human capital investments' (p. 920). In particular, they are more willing to invest in training, and there is a return to training.
- These findings are for white foreign-born male workers, and there are different findings for women and non-white migrants – as indeed there are for non-white native-born migrants. The findings also have to be understood as time-specific, being influenced by both the changing composition of migration over time, and labour-market conditions in the USA.

Source: after Chiswick (1978)

As emphasized earlier, human capital theories attracted considerable attention in the 1980s in the face of globalization, the intensification of competition, and the drive for enhanced competitiveness (Brown and Duguid 1991). Reich (1991), in particular, argued that there had been paradigm shifts in international competition, which meant that human capital had become not only a major but the sole long-term source of competitive advantage among more developed economies. He argued that labour forces would become polarized between those who did and those who did not have the skills to compete successfully in international labour markets. The contested roles of international migration, and the politics of migration policy, can be understood against this background of highly polarized and uneven welfare impacts across and within economies. This also brings us to the notion of skills, which are at the core of human capital theories.

Skills: an illusive concept

There has been a remarkable growth of research on skilled and highly-skilled labour migration since the late 1980s, reflecting both changes in the scale of such flows and the policy debate about the potential contribution of international migration to national stocks of human capital, and the knowledge base of the economy (see below). A number of researchers have commented on the prevalence of migrants in particular areas of science and R&D, especially in the USA (for example, Regets 2001) and Europe (for example, Salt and Findlay 1989; Salt and Ford 1993). There has also been a growing literature on international mobility and transnational companies (for example, Beaverstock 2005). These debates have spilled over into national and international policies about the role of skilled workers in economic development (OECD 2002), and we return to this theme in Chapter 5.

Such research – and indeed policy – has been highly selective in terms of gender (Kofman and Raghuram 2005), even among the highly skilled. There has been a strong focus on science and IT jobs that tend to be male-dominated, ignoring skilled occupations such as nursing which are dominated by women (see Winkelmann-Gleed 2006). Research has also been highly selective in relation to jobs within the skilled and highly-skilled categories. For example, Mahroum (1999, 2000) identified a five-fold categorization of mobility: managers and executives; engineers and technicians; academics and scientists; entrepreneurs; and students. Because of differences in migration duration, job content, and the organizations within which they work/study, these groups have different experiences of learning and knowledge transactions. The dominant focus has been on business professionals and senior managers, especially in context of intra-company transfers, so that the diversity of skilled migration has not been captured in existing research (Nagel 2005). Even though there is now an emerging body of research on scientists and entrepreneurs, there are other jobs – typically, technicians and engineers – whose engagement in international migration is still relatively little understood (but see Millar and Salt 2008).

There has been an even greater failure to examine in detail the skills of so-called 'unskilled and semi-skilled' migrants (Williams 2006a). There is of course considerable research on unskilled migrants, but this tends to focus more on social or mobility issues than on skills. Neither unskilled migrants, nor migrants in unskilled jobs, lack skills as such, rather they are to varying degrees employed in jobs where they can not (fully) utilize particular skills. In many cases, their status or their jobs are socially constructed as unskilled, whereas they may utilize high levels of other skills; this is often related to the gendering of migrants' employment (Raghuram 2004). Even migrants working in routine jobs – such as fast-food outlets – require skills, and this is an area that is still poorly understood.

In addition to the selectiveness of the literature, there is also the challenge posed by the illusiveness of the concept of skills: as Noon and Blyton (1997: 78) argue, 'skill is a definitional minefield'. One useful definition is provided by the *Concise Oxford Dictionary* (1976: 1071): 'Skill expertise is practised ability, facility in action or in doing or to do something'. Notwithstanding such definitions, several issues relating to skills have not really been resolved, and sometimes have barely been addressed in the migration literature. First, the definitions of skills – because these are central in the distribution of income, power and status – are socially constructed and politically negotiated. They are highly gendered, racialized and ageist, which has particular significance for national migration policies which increasingly discriminate on the basis of the highly contested notion of 'skilled vs unskilled' workers. Second, as Cockburn (1983) reminds us, there are three different perspectives on skill; whether this refers to skills located in individuals, to the skills demanded by the nature of employment or the skills required to perform a particular job. Indeed, much of the negotiation of definitions rests on the tensions between these different perspectives. Third, the measurement of skills is notoriously difficult (Bradley *et al.* 2000), and this is a perennial difficulty faced by human capital researchers.

The closely interwoven literatures on skills and human capital have both tended, traditionally, to focus on the technical aspects of skills as something that can be taught and assessed via formal means. This offers advantages for both policy makers (seeking measurable policy outcomes) and economists seeking to quantify and model the impacts of, and decision making relating to, international migration. Not surprisingly, the main measures of skills have tended to be selective (Keep *et al.* 2002: 11): years of schooling, wage levels, formal qualifications, occupations, courses and more formalized types of training. The problem with this approach is that (Brown 2001: 25) 'the obsession with measurable outcomes has led them . . . to ignore the process of skill formation', and to focus on 'formal episodes and ignore informal training and development'. The skills literature has also used employers' subjective assessments of whether they face skills gaps, as in the National Employers Skills Survey in the UK, which tend to reinforce the view that even where workers have the required formal paper qualifications, they may lack generic communication or problem-solving skills. In short, the measurement of skills has understandably had to rely on a limited range of surrogate measures of actual skills, particularly where the aim has been to produce national, regional or sectoral comparisons.

Language capital: the individual return from investing in language learning

Language skills feature strongly in the human-capital literature, and explain some of the initial wage differentials between migrants and native workers. Language is a key not only to being able to utilize fully the generic skills that migrants possess, but also to acquiring the nationally-specific skills that complement these in the destination labour market.

There are a number of studies of the language fluency of migrant workers. For example, a survey by Ruhs *et al.* (2006: 36) of East European migrants in the UK estimated that only 16.4 per cent considered that they were fluent, and a further 48 per cent that they had adequate English-language proficiency. In contrast, 26.1 per cent only had basic language skills, and 9.4 per cent stated that they had no language proficiency. Of course, the real meaning of these data depends on the distribution of length of residence in the UK, and the accuracy of such self-assessments, but the data do illustrate the range of migrant language skills.

Language is not always a barrier for migrants – for example, for migrants from the Spanish-speaking countries of Latin America working in Spain, or for Australian migrants working in the UK. However, even for these migrants, there can be other less obvious barriers associated with language and the way in which it is bound up with cultural meanings. As Elkjaer (2003: 43), argues:

> Language is according to social learning theory a central element of any process of learning since language is conceived as the main way of acting in contemporary organizations. Language is however not merely a medium of knowledge transmission. Language is the medium of culture and as such it constitutes a crucial element of the process of learning, when the latter is conceived as the result of interaction among individuals in a specific occupation, and organizational culture.

A similar point is made by Blackler *et al.* (1998: 75) who argue that 'Language does not passively mirror the world, rather speech is a practical act that shapes and negotiates meanings'. When you learn language it is of course socially constructed so that, for example, English has many more words for business practices than Slovak, Inuit people have many more words or expressions for snow than English. Therefore, in learning a language you are also learning about the social and cultural environment of the places where the language is used.

Although the cultural nuances of language are not directly addressed in human capital theories, researchers in this tradition have provided considerable evidence of the relationships between language skills, wages and employment. For example, Shields and Price (2003) reviewed the evidence for the UK and concluded that English-language fluency significantly enhanced the probability of being in employment by about a fifth to a quarter, after controlling for socio-demographic differences. Not surprisingly, the income premium to learning a language is greatest for those with the lowest fluency rates initially. Similarly, Dustmann and Fabbri (2002) estimated that fluency in English increased employment probability and earnings for migrants in the UK. Duvander (2001) provides an overview of similar evidence for Sweden, while Dustmann (1999) provides further illustration of the return to language capital in Germany (see Box 2.4).

There are a number of determinants of language skills and language learning (Chiswick *et al.* 2004), other than duration of migration and education. These include age: older migrants tend be less proficient in the language of the destination. Also significant are 'linguistic distance' between the immigrant's first language and the destination language, the intention of staying versus returning, and the degree of residential/community integration with migrants from the same national/ethnic group as opposed to the indigenous population. Interaction between spouses and between them and their children can also influence language fluency. This is expressed by Chiswick *et al.* (2004: 34) in these terms: 'If language learning takes place in the home, there is a spillover effect from one family member's investment in language training, namely, the improved language skills of other family members'.

Not surprisingly, immigrants who live in tightly-bounded ethnic enclaves, with few opportunities to practise the language of the host community, or to venture outside the 'security' of the home and neighbourhood, are also likely to have less well-developed foreign-language competence (Chiswick *et al.* 2004; Tomlinson and Egan 2002). For similar reasons, intermarriage between immigrants and non-migrants enhances language learning and communication skills (Chiswick and Miller 1995), again with spillovers into employment. Inter-marriage also facilitates the acquisition of country-specific customs, and knowledge of local labour markets (Meng and Gregory 2005), that is, nationally-specific human capital.

While broad measures of language capital or fluency can be obtained from surrogate measures, such as length of course taken, or from self-assessments, it is difficult to obtain a really accurate picture of this because the use of language is socially situated. In some jobs, a limited functional fluency may be enough, if skills other than communication are central to the work. But where communication is central – as in many professional and managerial posts – then migrants may find that it takes a long time to understand the

---

Box 2.4 **The return to language capital in Germany**

The classic human-capital perspective suggests that immigrants tend to adapt to their host countries via accumulating human capital. A critical element of human capital is fluency in the host country's language, which mediates their integration into that country's labour market. Language fluency is likely to bring higher returns in a host country than in the country of origin. The longer the migrant stays in a host country, the higher the expected returns to language capital. However, the incentives to learn language skills may be different for permanent and temporary migrants. Migrants who intend to stay for a longer time, or 'permanently', would be expected to invest more in acquiring language capital.

Dustman tested a model of investment by temporary migrants in language capital for a sample of 729 migrants who came to Germany, mainly between 1955 and 1979, and mainly from Turkey, Italy, Greece, Spain and the former Yugoslavia. The data was drawn from the German Socio-Economic Panel data set. A regression model showed that language fluency was negatively and significantly affected by the migrant's return propensity.

- Total duration of stay in the host country had a significant and positive effect on language expertise. The model showed that an increase in the length of stay by 10 years raised the probability of being fluent in German by 5 per cent.
- Migrants intending to stay permanently in Germany also had a 10 per cent higher probability of being fluent in German than migrants who intended to return.

Does fluent command of a host country's language impact on employment and wages? Dustmann and Fabbri (2002) addressed this, using data from two UK surveys (the Fourth National Survey on Ethnic Minorities and the Family and Working Lives Survey). They estimated that fluency in English increased employment probabilities by about 22 per cent and earning by 18–20 per cent.

Source: after Dustmann (1999); Dustman and Fabbri (2002)

---

nuanced use of language. Many migrants may never attain these 'upper reaches' of language use, because there is often a levelling-off effect (Taylor and Osland 2003): the costs of continued language improvement may be significant, leading migrants to cease learning when they have obtained a satisfactory rather than a maximum level of skill.

The brain redistribution debate: aggregate outcomes of the migration of human capital

There is an extensive debate about the aggregate net welfare effects of the international migration of (skilled) workers. Much of this is couched in terms of what we have termed the 'brain distribution' debate (Williams and Baláž 2005a). The origins of this often compelling, but still empirically thin discussion, lies in the seminal work of Grubel and Scott (1977), and Berry and Soligo (1969) but was significantly advanced and popularized by the work of Bhagwati and colleagues in particular (for example, Bhagwati and Hamada 1973). They crystallized the notion of international migration – seen as being mainly from Less Developed Countries (LDCs) to more developed ones – as an issue of brain drain versus brain gain. This was a limited view of the nature of migration, even at that time, and it has been overtaken subsequently by the growing importance of temporary migration, return migration, serial migration and circulation, as well as recognition of the importance of diasporic networks of skilled workers. Lowell and Findlay (2002) summarize some of the key perspectives in this debate, which we have represented schematically in Figure 2.2, with minor additions.

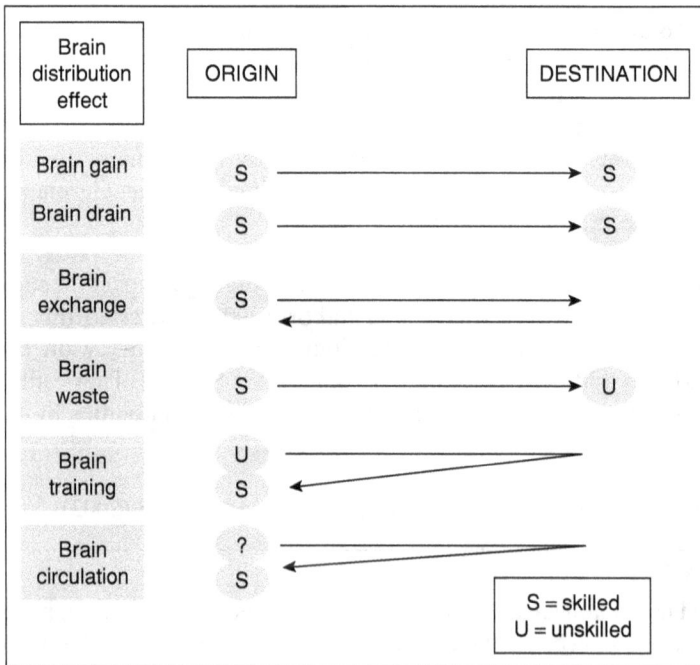

Brain distribution effect	ORIGIN	DESTINATION
Brain gain	S	⟶ S
Brain drain	S	⟶ S
Brain exchange	S	⟶ ⟵
Brain waste	S	⟶ U
Brain training	U / S	⟵
Brain circulation	? / S	⟵

S = skilled
U = unskilled

Figure 2.2 Schematic representation of 'brain distribution' perspectives
Source: authors

Essentially these typologies are based on the transfer and use made of human capital and skills by migrants and returned migrants in host and destination counties. The key issues are the levels of skills of the migrants, the selectiveness of out-migration, and the potential for under-utilizing skills either in the country of origin or the destination. Each proposed type of redistribution has different implications for sending and receiving countries, depending on the duration and the type of migration.

Straubhaar (2000) provides a useful guide as to why the redistribution of human capital is so important. The key factor is that migration creates positive external economies of scale for the destination countries, while there are also significant knowledge overspills in particular localities in the destinations. As a result, the brain drain restricts growth in the sending countries, while brain gain stimulates growth in the receiving country. Moreover, this develops into a vicious circle. LDCs lose skilled people which deters inward investment, so that average productivity levels remain low, reinforcing the incentive for highly-skilled workers to migrate. This provides a basic understanding of the brain-drain and brain-gain arguments. However, one of the problems with much of this literature is that it fails to take into account the quality of the learning and knowledge transfer environment in the destination country. For a number of reasons, migrants may actually take a job in the destination which, although it pays a higher wage than they had received previously, is actually unskilled. In this case, the net aggregate effect is termed 'brain waste'.

So far, these models envisage migration as a relatively simple one-way flow, from less to more developed countries. It does not recognize two other possibilities. First, that there may be relatively balanced flows between countries – typically between more developed countries – the aggregate effects of which are 'brain exchange'. And second, there has been a shift from brain drain to brain circulation (Gaillard and Gaillard 1997), which recognizes return migration and circulation. Brain training – involving student migration – is a special case of this, where the migrants do return to their home country. This last point emphasizes that all these models are static, and that the positions of individuals, and of entire economies, may change over time.

In their simplest forms, the models of brain drain, brain gain, and brain waste suggest that there are clear losers and winners in this redistribution of welfare. The losers are the sending countries which paid for the early care and education of the migrants (that is, invested in their human capital), while the winners are the developing countries which have effectively off-loaded these investment costs to the LDCs. The New World, especially the USA, was the big winner in the nineteenth and the early decades of the twentieth century, but northern and central Europe, and to a limited extent Japan, also became winners in the latter half of the twentieth century. This is of course an oversimplification, as Mahroum *et al.* (2006: 27) states:

> The debate about the benefits and losses of skilled migration (seen as a
> 'brain drain' for source countries and 'brain gain' for receiving countries)

reveals that this is a complex subject, and one in which all the costs and benefits to various stakeholders have not yet been carefully studied.

There are several layers of complexity that need to be disentangled. First, do the migrants send remittances to their home countries and do these outweigh the direct loss of human capital (Goladfarb *et al.* 1984)? Remittances are also advantageous because they are less volatile than capital flows. Second, will the migrants return while still economically active, and contribute to their home economies – whether after a brief or a long period of emigration – or will they return to inactivity and retirement? Third, can the home economies benefit from the knowledge and capital accumulated by the migrants even if they do not return home? – this is associated with more recent attempts to utilize the resources within diasporic networks (Meyer and Brown 1999). Fourth, there is also an argument (Mountford 1994) that brain drain might, seemingly paradoxically, actually lead to increased productivity in the sending economy – through providing an incentive for other workers to invest in training. This has led to a view (Stark 2003) that there may be an optimal level of skilled migration, if it does lead to greater personal investment in training in the sending countries. However, there is also a counter-argument that it may lead to indigenous workers in some LDCs being unwilling to take up lower paid or menial jobs. As with much of this debate, that there are still relatively few detailed empirical studies of these effects (Findlay 2006) and it is difficult to know for certain whether international migration leads to anything more than 'brain strain' (Lowell *et al.* 2004).

The empirical evidence on the distribution of migrants confirms the complexity of these brain distribution effects. Abella (2006: 16–17) provides a relatively comprehensive review of the educational attainment of migrants as reported in several population censuses taken circa 2000. In the 29 OECD countries, an estimated 46 per cent of the foreign-born populations originated from another developed country – providing strong evidence for population exchange. Of these, some 6.4 million or 17.6 per cent of the total are skilled (having tertiary-level education), which provides evidence of brain exchange. The main origin countries outside the OECD are the former USSR, former Yugoslavia, India, the Philippines, China, Vietnam, Morocco and Puerto Rico. The former USSR had the largest graduate expatriate community (1.3 million), followed by India (1 million). In these cases, there is brain drain, brain waste, or brain circulation, but there is lack of firm evidence as to exactly which outcome prevails. However, whereas migrants from Developed Countries tend to be temporary, those from the LDCs tend to be long-term or permanent, suggesting that brain drain may well be the best descriptor. In any case, the real cost of such migration can be formidable and is underlined by the stark statistics that Jamaica has to train five doctors, and Grenada 22 doctors, to keep just one doctor in the country (Lapper 2007: 7). See also Box 2.5 for a discussion of the costs of brain drain.

Box 2.5 **The cost of brain drain**

Carrington and Detragiache (1998) estimated the stock figure of the total brain drain from Less Developed Countries (LDCs) to OECD member countries as approximately 12.9 million people (of whom some 7 million went to the USA). The poorest countries in the world have been most affected by the brain drain of educated labour. For example, there has been a striking exodus of doctors and engineers from Africa in the past two decades. Between 1986 and 1995, for example, 61 per cent of graduates from one Ghanaian medical school left the country. Of over 600 medical graduates trained between 1977 and 2000 in Lusaka, Zambia, only 50 were still working in the Zambian public-sector health service in 2000. In 1978 Sudan lost 17 per cent of its doctors, 20 per cent of its university lecturers, 30 per cent of its engineers and 45 per cent of its surveyors (Schrecker and Labonte 2004).

The exodus was caused by a mixture of economic and political factors, of which wage differences were usually the most significant. For example, a nurse in the USA can expect to earn $3,000–4,000 per month, compared with just $300–800 per month for a doctor in the Philippines which, even adjusting for differences in purchasing power, represents a dramatic difference. But wage migration is a relationship not between poor and developed countries, but also from very poor to less poor developing countries. The average salary for junior doctors in Lesotho, Namibia, and South Africa in 1999 was five times or more the level in Zambia or Ghana, and 20 times the level in Sierra Leone.

The impacts of brain drain from Africa are particularly visible in two areas: deteriorating public health-service systems, and direct financial losses:

- In 1998, vacancy rates for doctors in the public health sector were estimated at 26 per cent in Namibia, 36 per cent in Malawi, and 43 per cent in Ghana. In contrast, Canada was a major destination for doctors from southern Africa, In Saskatchewan, for instance, the Canadian province most heavily reliant on foreign medical graduates, almost one in five of the province's doctors in 2001 had obtained their first medical degree in South Africa. These 260 physicians represented the equivalent of 5 years' output from the University of Saskatchewan's medical school (Schrecker and Labonte 2004).
- Training a medical doctor costs $60,000 and training a paramedic costs $12,000 in developing countries (plus the costs of primary and secondary education). These countries are, effectively, subsidizing the OECD countries to the order of an estimated $500 million per year, just in terms of training medical staff. The savings for the more developed countries are immense. UNCTAD, for example, estimates that the per capita training costs saved are of the order

of $184,000 for each professional aged 24–35. Given that there are an estimated 3 million professionals trained in the LDCs who are now working in the 27 OECD states, this suggests a staggering $552 billion in savings for the latter (IOM 2005).

Source: after Carrington and Detragiache (1998); Schrecker and Labonte (2004); IOM (2005).

Although neither the theoretical nor the empirical evidence as to the net aggregate effects of 'brain redistribution' are unequivocal, there is no doubt that as Gamlen (2005) argues 'National territorial boundaries no longer delineate the extent of the nation's human capital' and that this has profound implications in terms of winners and losers.

From skills to competences

The previous section reviewed some of the key ideas in human capital theories, one of the most powerful theoretical perspectives in migration research, as evidenced by its penetration of both research and policy agendas. These theories also have limitations, including the difficulties of measuring skills (Auriol and Sexton 2002). As Brown argues, 'Human capital theorists have either ignored the importance of "key" skills because they are difficult to quantify or treated them as technical competences to be taught, learnt and assessed in a formal way' (Brown 2001: 24).

In recognition of these difficulties, there have been changes in how skills are conceptualized. 'Skill' is now seen in broader terms than in the past, when it was often associated with manual craft workers or technology-oriented tasks (Payne 2000). Reich (1991) sought to address this deficiency by identifying three types of skills: technical (involving high levels of symbolic manipulation), routine skills (repetitive work), and social skills (which facilitate communication and social interaction). To some extent, the first two broadly equate with the notions of 'skilled' and 'semi-' or 'unskilled' workers, although they are more precise terms, while social skills recognizes an array of very different abilities. This equates with the broader debate in the skills literature about 'soft' and 'interpersonal skills', as summarized by Payne (2000: 354): 'skill has expanded almost exponentially to include a veritable galaxy of "soft", "generic", "transferable", "social" and "interactional" skills, frequently indistinguishable from personal characteristics, behaviours and attitudes, which in the past would rarely have been conceived of as skills'. In addition, issues of motivation and attitude can be seen in terms of cultural issues, and 'emotional intelligence' (Goleman 1998). Researchers are still coming to terms with this different way of understanding skills, most notably through the notion of competences. Migration research, by and large, has lagged behind this paradigmatic shift.

One of the driving forces behind the shift from skills to competences in the generic literature has been the realization (by policy makers and employers) that 'qualifications may serve as a sorting mechanism in recruitment rather than an indicator of productive potential' (Rainbird 2000: 185). There has therefore been an attempt to produce a more nuanced understanding of the abilities and capabilities of individual workers, by drawing on the finer-grained concept of competences. This is an approach which emphasizes a range of abilities and experiences, acquired in various ways and in various places, constituting accumulated formal and informal learning over the life course. Keep *et al.* (2002: 12) summarize the shift in conceptualization from skills to competences:

> Twenty years ago skill tended to be thought of in terms of theoretical knowledge, intellectual abilities (e.g. reasoning), and various forms of manual dexterity, or a combination of these elements, though even then research suggested that employers' conceptions of what made someone 'skilled' were malleable. . . . More recently, the notion of skill has acquired a number of additional aspects. These include generic skills or competences (such as the ability to work in teams, the ability to communicate, the ability to solve problems, etc.); personal attributes (such as leadership, the ability to be easily motivated and to motivate others, politeness, a willingness to compromise, and positive attitudes towards change and authority); and appearance (what some have dubbed aesthetic labour).

Evans (2002) provides a useful typology of competences which crystallizes some of this thinking in his so-called 'starfish' model:

- content-related and practical competences (e.g. willingness to carry out a variety of duties);
- competences related to attitudes and values (e.g. responsibility, or reliability);
- learning competences (e.g. openness to learning, or perceptiveness);
- methodological competences (e.g. networking skills or ability to handle multiple tasks); and
- social and interpersonal competences (e.g. communication skills or awareness of others' viewpoints).

The learning competence is particularly important. Brown (2001: 15) emphasizes the importance of the ability to 'learn how to learn' in terms of the increasing pace of change in labour markets: 'The short life cycle of knowledge and skills makes the ability to grasp new information and acquire new skills an inherent feature of skill formation in post-industrial societies'. This has particular resonance in relation to migration, which can be understood not only as a response to these changes but also as constituting a learning experience.

This finer-grained approach of competences does partly allow the researcher to look beyond politically-negotiated or socially-constructed definitions of skills, although any such definitions in this realm are necessarily social and political constructions. And there is also 'a distinctive whiff of elitism' in relation to some of these softer skills (Payne 2000: 363) especially attributes such as social and interpersonal competences. Nevertheless, they do provide a conceptual advance that opens up a fuller perspective on the range of skills, that looks beyond the technical, and recognizes that both 'skilled' and 'unskilled' migrants should be understood as idealized positions on a continuum of competences, which they possess to different degrees and in different combinations.

The shift from skills to competences is not, however, unproblematic, and has attracted criticism, not least because, as Ainley (1993: 357) argues:

> 'Skills' formerly understood by many as complex *social* processes, were now de-contextualized and de-constructed into finite, isolable 'competences' to be located as the property of the individual, who then carried them, luggage-like, from job to job.

It is a trenchant criticism and one which has particular resonance for migration studies, where the 'luggage', as Ainley terms it, has to be carried not only from job to job, but also across international borders.

Notwithstanding this reservation, the notion of competences is starting to infiltrate migration research, as Berset and Crevoisier (2006: 78) recognize:

> Today, competences have become a much more central issue. As innovation becomes more crucial, access to a very broad and diversified international labour market appears as an important trump card. The 'quality' of the individuals is becoming decisive, as is the capacity of companies to capture, mobilise and orient these competences. In such a context, the circulation of competences via migration plays a much more central role than it did in the past.

It is, of course, far more difficult to measure or assess such competences, than relatively easily quantifiable traditional indicators of skills, such as years of schooling. However, some of these competences have been noted and commented on by researchers in selective areas of migration studies. First, there are links to the literature on networking (Meyer 2001; Zhou and Tseng 2002), although this focuses on how networks are used rather than on how the competence to network (what Evans 2002 refers to as a 'methodological competence') is developed, is transferable and is commodified in different places.

Second, there has also been a focus on the role of international migration in developing networks, and acquiring knowledge of different national business cultures, among professionals in the advanced business services. This stems from the research on transnational enterprises, the management of intra-company knowledge flows, and securing 'nationally-specific' knowledge of particular markets, customers and business practices as a key component of competitiveness and strategies for expansion (McCall 1997). But it is also a process whereby young professionals and managers acquire particular competences that equip them for their roles in transnational corporations. Geographers have been particularly interested in the place-specific nature of such knowledge transactions, and in the ways in which local and distanciated networks interact (Beaverstock 2002; Yeoh and Willis 2005).

Third, the authors (Williams and Baláž 2005a) provide one of the few explicit engagements with the notion of competences in the migration literature, applying Evans' typology to understanding the experiences of contrasting groups of Slovak returned migrants from the UK. Their research demonstrated that even where migrants had relatively 'unskilled' jobs in the UK, such as au pairs, they were able to acquire particular competences, in terms of methodology (networking), learning (openness), attitudes and values (self-confidence), and communication (English-language skills). These were commodifiable in a range of jobs after returning to Slovakia. Finally, Matthews and Ruhs (2007), although not specifically writing about competences, provide further illustrations of the role these play in the hospitality sector (Box 2.6).

Towards a knowledge perspective

Advocating a knowledge perspective

While competences offer an advance on the notion of skill, this is also a problematic conceptualization in some respects. Whereas human capital theories recognized that there were both generic and nationally-specific skills, this is seemingly denied by the notion of competences which can be transported with relative ease from job to job (Ainley 1993). In contrast, this book contends that a knowledge perspective offers four main advantages for migration researchers.

First, and as will be explained in this section, it is possible to understand knowledge as having both generic and socially-situated strands. Polanyi's (1966) classic distinction between tacit and explicit knowledge has been elaborated over time, and Blackler (2002) identified four main types of tacit knowledge: embodied, embrained, encultured and embedded. The first two are relatively intrinsic to the individual, while the latter two are more reliant on shared meanings. Therefore, unlike competences, knowledge cannot be transferred across borders in an unproblematic manner. Rather, knowledge has elements that are place- and or culture-specific, as well as being embedded in specific institutions.

Second, it places the personalized knowledge of individuals, equated by Polanyi (1975) to tacit knowledge, in context of total knowledge in society. Total knowledge includes the collective knowledge possessed by organizations

***Box 2.6* Are you being served? Competences and migrant employment in the UK hospitality sector**

According to the Labour Force Survey, 22 per cent of workers in the UK hospitality industry were migrants in 2006. Of these, two-thirds were from non-EU countries, but the share from the new Eastern European member states was increasing rapidly. There are three main reasons why employers employ migrants in this sector, and the third of these approximates to the notion of competence.

- Labour market segmentation is a well established strategy for reducing labour costs. This is based on migrants being more willing than non-migrants to accept lower wages and difficult working conditions because their main reference point is their home country.
- Employers seek to reconcile the contradictory demands of wanting both a flexible labour force, and retaining workers, by recruiting migrants over whom they can exert considerable power by various means, including the visa system and providing tied accommodation.
- Employers purchase potential rather than actual quantities of labour, in employment contracts, as the latter is dependent to some extent on the efforts and attitudes of individual workers. They are therefore faced by indeterminancy which they seek to reduce by recruiting workers with 'good attitudes'. They also seek to recruit workers with what they consider to be appropriate social, tacit and aesthetic qualities. As employers find it difficult to assess such competences, they often base recruitment on stereotypical assumptions about particular migrant groups (Waldinger and Lichter 2003).

Source: after Matthews and Ruhs (2007)

and individuals, and codified or explicit knowledge, including technologies (Polanyi 1966). This is an important point because it potentially allows migration researchers to link their work to that of economists, economic geographers and economic sociologists and engage in debates relating to the knowledge economy, innovation and the management of knowledge. Migration can be seen as integral rather than as extraneous input into these literatures.

Third, a knowledge perspective also provides a way to avoid the intellectual trap posed by the simplistic dichotomy between skilled and unskilled migrants. However, a note of caution is required here because much of the literature on knowledge in the economy has recently been dominated by the notion of the knowledge-based economy. This is, without doubt, an over-hyped concept. Lloyd and Payne (2003: 87), for example, refer to 'the constant drip-feed of rhetoric surrounding the new knowledge economy, as supplied by policy makers and a succession of popular texts'. It is, at most, an emerging dimension of the economy, and there is still ample evidence of the persistence of Taylorist and Fordist production methods, and deskilling, in many sectors. More importantly, for our purpose, it tends to reify particular types of skilled workers, in the same way as the migration literature has reified the skilled migrant. A more realistic generic picture is probably presented by Warhurst and Thompson (1998: 8) who argue that there is continuing variation in workplace practices, alongside a general trend towards the intensification of work. It is, therefore, more useful to follow Thompson *et al.*'s (2001) notion that all workers are knowledgeable, and – as argued in Chapter 1 – this leads us to the notion of the knowledgeable migrant, irrespective of the jobs that he or she held in either the country of origin or destination.

Fourth, the human-capital perspective assumes that migrants' wages will rise over time as they acquire nationally-specific knowledge or skills in the destination. This is an inherently static view of the economy, because it effectively considers migrants to be 'replacement knowledge bearers', who necessarily have to adjust to the knowledge framework in the destination in order to fulfil their roles more effectively. This implicitly denies them the role of bringing with them distinctive knowledge which may be valued precisely because it is different to the prevailing nationally-specific knowledge system.

Having set out the case for a knowledge perspective on migration, we proceed to set out a theoretical framework for understanding migration as a channel for learning and knowledge transfer.

The transferability of knowledge via international migration

Hodkinson *et al.* (2004: 11) comment on the differential transferability of skills and knowledge between different working environments:

> prior abilities are important in negotiating changes of work and learning environments. These are not decontextualised 'transferable skills' but abilities that have structural and referential features. Their structural features may be carried (tacitly) between environments but they have to be situated, underpinned by domain-specific knowledge and developed through social interaction within the culture and context of the work environment.

The distinction between structural (transferable), and domain-specific knowledge is useful, not least because it broadly accords with the differentiation between embrained and embodied knowledge as opposed to encultured and embedded knowledge, which we develop further below. However, as Evans and Rainbird (2002: 24) note, understanding 'the processes by which skills are transformed from one setting into another' is limited, and this is especially so for international migrants.

The key to such transmissions is that knowledge and learning are relational, so that transfers between individuals in the same setting (for example, a company and country), let alone transfers between settings, are perhaps better thought of as *translation*. Czarniawska (2001: 126) elegantly captures this: 'It is people whether regarded as users or as creators, who energize an idea every time they translate it for their own or somebody else's use. Watching ideas travel . . . we observe a process of translation'. This process of translation modifies all the agents involved: the individual translators, and the translated knowledge. The notion of translation takes us beyond simplistic ideas about transferring immutable knowledge, and leads to consideration of knowledge creation. There is a very fine line between knowledge translation and creation. Migrants bring knowledge with them to a new setting, where it may be integrated with other knowledge through participation in various formal and informal practices, not only within but also outside of employing organizations. As a result, their knowledge can be described as 'having been expanded, modified, or even transformed' (Eraut 2000: 27). At some point, therefore, knowledge translation (approximating to 'expanded' and perhaps 'modified') becomes knowledge creation (approximating to 'transformed').

In order to ground the following discussion more firmly in an understanding of the nature of knowledge, we need to revisit the notion of tacit knowledge, and for this purpose draw substantially on Williams (2006a, 2007a, 2007b). Polanyi's (1958; 1966) seminal work, distinguishing between tacit and explicit knowledge, is the obvious starting point for considering different types of knowledge. This essentialized tacit knowledge as being person and context specific, and this can be paraphrased as 'knowing more than can be expressed verbally'. In contrast, explicit knowledge is transmittable in formal and systematic ways (via manuals, databases etc.). This dichotomy has been extended by a number of writers, notably by Nonaka and Takeuchi (1995), who identified four types of knowledge transfer, involving different combinations of tacit and explicit knowledge. Subsequently, other writers have developed finer-grained conceptualizations of knowledge, for example Yang (2003) who distinguishes between explicit, implicit and emancipatory knowledge. But this paper adopts Blackler's (2002) typology, which draws especially on his own earlier work and that of Zuboff (1988), Berger and Luckmann (1966), and Brown and Duguid (1991). This typology is particularly useful for understanding how international migration mediates particular forms of knowledge transfer.

- *Embrained* knowledge is dependent on conceptual skills and cognitive abilities, which allow recognition of underlying patterns, and reflection on these. The individual mindset is a key influence on learning.
- *Embodied* knowledge results from experiences of physical presence (for example, via project work). This is practical thinking rooted in specific contexts, physical presence, sentient and sensory information, and learning in doing.

- *Encultured* knowledge emphasizes that meanings are shared understandings, arising from socialization and acculturation. Language, stories, sociality and metaphors are mainsprings of knowledge.
- *Embedded* knowledge is embedded in contextual factors and is not objectively pre-given. Moreover, shared knowledge is generated in different language systems, (organizational) cultures and (work) groups.
- *Encoded* knowledge is embedded in signs and symbols to be found in traditional forms such as books, manuals, codes of practice, and websites.

The different types of tacit knowledge are exemplified in Box 2.7 which draws on our work on Slovak returned migrants working in the health sector.

***Box 2.7* Knowledge transactions and returned migrants: the Slovak health system**

This study of returnee doctors in Slovakia in 2006 was based on in-depth interviews with 19 doctors and five managers. They had spent relatively short periods abroad: six had spent three–four months and another five had spent up to one year, while eight had spent longer periods up to a maximum of four years. Their experiences ranged from advanced study, to periods of employment, although in medicine these are often blurred. The study also covered a considerable time-span; whereas pre-1989 migrants, under state socialism, had largely been limited in their choice of destinations, for example to other eastern bloc countries, the main destinations after this date had been Germany, Austria, Switzerland and the Netherlands.

Both managers and doctors agreed that international mobility had been a source of significant and distinctive learning and, to a lesser extent, an opportunity to transfer some of their tacit knowledge to fellow doctors in the host insitutions. Embrained and especially embodied knowledge were highly prized in their learning experiences, but so too were the perspectives provide by the encultured and embedded knowledge of different health systems. For some doctors, the acquisition of advanced specialized knowledge – much of it learning by practice or learning by observation – was the main return that they had secured from mobility, but for others it was learning about what one doctor termed, 'different philosophies' of health care. Most of the barriers to learning abroad were overcome after initial difficulties, for example, in terms of language competence. But some doctors suffered persistent obstacles related to their positions as outsiders.

Their mobility experiences did generally contribute positively to their work after returning to Slovakia, although a small number considered that they had been marginalized. They had also generally been able to transfer some of the knowledge they had acquired abroad to

---

Slovak colleagues. However, there were differences in the extent to which organizations and some individuals were open to the transfer of such external knowledge. Some of the returnees complained about the obstacles faced in introducing new ideas into systems of strongly embedded knowledge within hospitals where power was often strongly centralized. But knowledge transfer had been possible, whether in codified (articles in journals, or written reports) or tacit forms (via demonstration, or in conversations).

This study provides illustrations of each of the main types of tacit knowledge identified by Blackler (2002), as well as of encoded knowledge.

- Encoded knowledge was represented by the books they brought back with them to fill the deficiency which had existed until late in the 1990s. One doctor recalled: '*Back in the 1990s there was a lack of books, and there was no internet. If I bought a book then it cost 25,000 koruna which was my salary for a month, so my family would starve. What I used to do was go through the Current Contents, and find interesting articles. Then I would write to the authors asking for copies. . . . Now we have the internet so there are no real differences in medical knowledge*'. International mobility to acquire tacit knowledge however remained important despite the growth of electronic forms of communication and dissemination.
- Embrained knowledge was typified by learning about diagnoses, or identifying patterns of symptoms by working alongside leading international experts. For one doctor it was simply the sheer volume of cases that he had worked on which was critical: '*I cannot say I learnt anything really different there. . . . What really was new, were large numbers of patients and operations in one place. It allowed me to know a broader spectrum of patients and do far more operations than in Slovakia. . . . In Slovakia learning would have taken much longer.*'
- Mobility was considered to be particularly important for acquiring embodied knowledge. One returnee commented on his experiences in Switzerland: '*You observe similar operations, similar diagnoses and patients and similar solutions – but! – you see minor differences, which are not mentioned in books. You can see and learn some details, which are very useful to learn. And I really did. You can see it and imitate it, because this is a practical matter. They tell you this is something new and we do it in such and such way. You discuss it and remember.*'
- Encultured knowledge was particularly important in terms of understanding the different perspectives not only of professionals but also of patients: '*Western European patients know more about*

> *their diseases. In Slovakia, patients and relatives don't really com-
> municate with doctors. . . . Why is it different in Slovakia? It's not
> part of our training or work culture – or even of our national culture.
> Slovak patients are just different'.*
> • Embedded knowledge was expressed in terms of learning about
>   different systems of health care: '*It is good to go somewhere and
>   see how the different system operates. . . . When you return home, you
>   can compare procedures and think about whether you should change
>   your routines or not. Sometimes it pays, sometimes not'.*
>
> Source: after Williams and Baláž (2008b)

There has been increased recognition of these different types of knowledge because of the changing organization of work, notably greater emphasis on so-called 'soft' skills of communication, problem solving and creativity (Payne 2000). The issue here, however, is the transferability of particular types of knowledge via (international) migration. Encoded knowledge is, of course, the most mobile form. In contrast, tacit knowledge is inherently less mobile because it cannot be fully articulated through documented (i.e. codified), and possibly even through verbal, forms, but is learned through experience (Nonaka and Takeuchi 1995; Polanyi 1966). Consequently, given that much knowledge is tacit, 'whether knowledge spillovers are contained within a firm, or within a region, depends on the nature of mobility in that industry or region' (Almeida and Kogut 1999: 916).

The key question, however, is the relationship between knowledge transferability via migration and Blackler's (2002) typology of tacit knowledge. Embrained and embodied knowledge are necessarily indivisible from the individual, and so are fully transferable via corporeal mobility. Migrants who have cognitive skills that allow them to diagnose faults in computer language (embrained knowledge), or sensory and physical skills for restoring art works (embodied knowledge), can transfer these in their entirety across borders. Of course, there may be national differences in the types of embrained and embodied knowledge that workers have acquired in different places – stemming from differences in national educational and training systems, or from the particular industrial mix in the economy. The return to the migrants' knowledge therefore is conditioned by their migration between countries where there are different market values for embodied and embrained knowledge.

In contrast, encultured and embedded knowledge are grounded in particular places and institutions. In so far as these 'settings' are not transferable or replicable, then – at best – they are only partly transferable through corporeal mobility. A negative interpretation would emphasize that they are necessarily devalorized by migration – which is akin to the time–earnings function in human capital theories. However, this denies individuals the capacity for reflexivity and, by extension, for migrants to take with them *knowledge of* encultured and embedded knowledge, even if these are time- and place-specific and not totally transferable because they rest on shared understandings. Moreover, they may be able to draw on their previous embedded and encultured knowledge to provide a comparative perspective on the nature of encultured and embedded knowledge in different countries (Williams 2006a). They may, for example, be able to reflect on how organizational culture in their new employer influences production and productivity, compared to their last job. All mobile workers can reflect on organizational differences (especially on embedded knowledge), but international migrants bring an additional dimension to this, because of the significance of national boundaries in the distribution of encultured knowledge. In other words, international migrants may have a particular capacity for reflexivity that can be related to the Critical Reflection School of Action Learning. As summarized by Marsick and O'Neil (1999: 163), participants 'need to reflect on the assumptions and beliefs that shape practice. . . . Critical reflection can also go beyond the individual participant's underlying assumptions and can lead specifically to the examination of organizational norms'.

Migrants can transfer ideas about alternative work practices, products and especially markets, even if these require modification to fit culturally and organizationally different settings. In part, they carry such knowledge with them, but in part they carry the means to access such knowledge through their social networks which extend across international boundaries. Such linkages are particularly intensified in respect of transnational migrants (Zhou and Tseng 2002). This is not to deny that there can be considerable obstacles in seeking to apply reflexive encultured and embedded knowledge in different settings. For example, transferring embedded knowledge between organizations is problematic because it resides in an organization's inter-related systems of physical, human and organizational relationships (Empson 2001). Individual migrants have limited capacity to bring about changes in such institutions (see Chapter 6).

This reflexive capacity is associated with the role of knowledge brokering, in which international migrants can be particularly significant because of their potential to transfer 'uncommon knowledge'. As Bentley (1998: 157) argues:

> Relationships which are distributed across organisations, social groups and geographical areas connect us to a wider range of resources and help to broaden our horizons. From this perspective, it is the most surprising and unconventional relationships that are of most use. If our relationships mirror the formal external structures by which we organise our lives – school classes, tiers of management offices, families – then our access to information and resources is determined by these structures. Much of what we can learn from co-workers will already be common knowledge.

International migrants as potential knowledge brokers have a number of different roles to play. Following Tushman and Scanlan (1981), they can be boundary spanners who take care of one boundary at a time, 'roamers' who travel from place to place, and 'outposts' who bring back knowledge from 'the forefront'. Changes in the nature of international migration have increased the potential of migrants to perform all three of these roles. For example, there has been a growth of short-term migration (King 2002) and of transnational migration, both of which facilitate boundary spanning and roaming. Transnational migrants are especially likely to act as brokers, because they have access to embedded knowledge across international borders. And growing temporary migration (Ruhs 2005) means there is an increasing number of returnees who transfer knowledge from so-called outposts. This is not to argue that all or even most migrants are knowledge brokers or boundary spanners (Tushman and Scanlan 1981), but rather that their embedded and encultured knowledge may have positive value in some circumstances. It is not always a disadvantage for migrants to have different nationally specific knowledge to that in the destination, although relatively few migrants have the full range of skills to be effective in such roles.

One other caveat also needs to be noted. The central argument in this section has been that embodied and embrained knowledge are more easily transferable via migration than encultured and embedded knowledge. This however is to assume that all international migration is from countries of birth to new destinations, a denial of the growing importance of temporary and circulatory migration. Returned migrants obviously possess appropriate and nationally specific embedded and encultured knowledge, unless they have been abroad for exceptionally long periods. Similarly circulatory migrants will have acquired embedded and encultured knowledge in more than one place.

Knowledge can be transferred across space via many different channels, but migration involves a particular combination of embrained, embodied, encultured and embedded knowledge. Therefore, while this section has distinguished between different types of knowledge, one of the keys to their valorization is how they are combined. In this sense, all forms of knowledge are relational, and none are transferable by migrants without transforming their potential economic value. The question then is how corporeally mobile forms of knowledge, and 'knowledge of knowledge', are recombined with other forms of knowledge, in new settings which may be politically, culturally and organizationally different. We still know relatively little about this topic, but Box 2.7 provides some insights into such a perspective, as applied to returned migrant doctors in Slovakia.

We have argued here that international migration brings about face-to-face relationships that can facilitate tacit knowledge transactions. Of course, migration is only one means for achieving corporeal co-presence. Koser and Salt (1997: 299) argue that 'if physical mobility is required to make . . . expertise mobile, it may range in duration from a few hours in a workshop, to a few months on a placement, to long term migration'. Changes in technology, transport and knowledge management mean that different forms of mobility have become increasingly substitutable (Salt and Ford 1993: 27). Tacit knowledge can also be shared electronically, leading Amin (2002) to argue that localized and distanciated relationships may be interwoven. Despite these caveats, international migration is increasing in scale, and changing in composition, constituting an important, but little researched, channel of tacit knowledge transactions.

Labour market segmentation: double and triple labour markets

Before leaving this review of the theorization of the use of skills and knowledge, we need to address one other, rather different, theoretical perspective on the labour market experiences of international migrants. It is well known that migrants often tend to be concentrated in particular segments of the labour market. According to Castles and Miller (1993: 190), this is because of a combination of regulatory, institutional and cultural reasons:

> The process of labour market segmentation usually results from a combination of institutional racism and more diffuse attitudinal racism. This applies particularly in countries that recruit 'guest-workers' under legal and administrative rules that restrict foreign workers' rights in a discriminatory way. The legally vulnerable status of many foreign workers in turn fosters resentment against them on the part of citizen workers, who fear that their wages and conditions will be undermined. This may be combined with resentment of foreign workers for social and cultural reasons, leading to a dangerous spiral of racism. Such factors have profoundly affected trade unions and labour relations in most countries that have experienced labour immigration since 1945.

Castree *et al.* (2004: 56) have usefully placed labour market segmentation in a broader perspective, arguing that it is the outcome of three main features. First, segmentation of labour demand – employers have variable demands for workers in terms of both quantity and quality, and they use surrogate measures such as age and nationality when recruiting and allocating workers. Second, segmentation within workplaces, which is reinforced by the lack of career ladders in many jobs. Third, segmentation of the labour supply, as for example in the way in which gender divisions of labour in the home inform recruitment and working practices.

Labour market segmentation theory is particularly associated with Piore's Dual Labour Market Theory, which essentially understands labour markets as divided into primary and secondary sectors. The primary sector is characterized by higher wages, relative job security and career ladders providing at least some potential for upward mobility from many positions. The secondary sector, characterized by low qualified jobs, precariousness and informality is typically the destination of 'unskilled' migrants. There is also a relatively low level of mobility between the two sectors, and individuals (migrants) in the secondary sector tend to become entrapped within it. Skills are learnt on the job, and there is little or no formal training.

Although Piore's work has been invaluable for migration theorists, it has limitations, two of which we note here. First, it fails to recognize that there may also be segmentation of migrant workers within the primary sector. Findlay (2006) provides evidence that this has been the experience of many skilled migrants in the UK in recent years. Many of these migrants, particularly in the public sector, ended up in the jobs which were least popular with the native-born population, for example teaching jobs in inner-city schools, or some of the experientially less rewarding branches of the National Health Service, such as geriatric medicine. He concludes that 'One result of the UK's "employer-led" policies may therefore have been segregation within skilled labour markets of many immigrant staff' (Findlay 2006: 88).

Second, the theory has been criticized for ignoring the existence of enclave labour markets, where ethnic minority (in this case, we are concerned with migrants) employers employ mostly co-ethnic workers. There are a number of theories to explain this phenomenon, emphasizing either that this is an opportunity to utilize particular resources, or that it is a 'haven' for employers and employees in the face of disadvantages encountered in the main labour markets. We discuss these ideas further in Chapter 6. Here it is sufficient to note that Portes and Bach (1985) have responded to this critique with their 'triple labour theory' which identified a third major segment, the enclave labour market.

All three sectors provide different learning and knowledge transfer opportunities for migrants (McGovern 2007). The primary sector, with better-paid and more secure jobs, provides the best opportunities for formal training simply because employers will have a greater incentive to invest in these given their more stable workforces. The enclave labour market will provide least opportunity for learning from other national and ethnic workers, and for reaping the creativity dividend of diversity. Massey (1994: 720) provides a useful overview of the empirical findings (up to the early 1990s) and concludes that:

> The accumulated evidence thus appears to indicate that US labor markets are indeed segmented; that immigrants are selectively excluded from the primary labor market and found disproportionately in the secondary labor market, where they earn limited returns to education, skills, and experience; but that immigrant enclaves provide significant returns to human capital and an alternative mobility ladder for immigrants in some cities, especially for male entrepreneurs.

In terms of the enclave segment, the key question is whether this represents a stepping-stone or entrapment for migrant workers, a theme that we return to in Chapter 7. Beyond this, it is difficult – and probably highly contingent – to understand how the more informal opportunities for knowledge transfer and learning differ between these three sectors. Nationality and ethnicity intersect with age and gender, as well as with the features of migration (duration, timing or channels used). At one level it is undoubtedly true that there is a 'unique situation that pertains in each professional area' (Iredale 2001: 15), although this does not preclude us from trying to identify the most appropriate theoretical tools for understanding these diverse individual experiences.

The theorization of labour market segmentation has subsequently been significantly advanced by the work of Waldinger and Lichter (2003) in *How the Other Half Works*. Their starting point is a critique of the debate about the inevitability of deskilling vs upgrading in the labour market, and they conclude that both may occur simultaneously, with this providing the context for the insertion of migrants in the labour market. When filling so-called unskilled jobs, employers cannot rely on automation turning workers into mere automatons. Rather, all jobs require some proficiencies (what we have termed competences earlier in this chapter), including attitudes and motivations – it may be the personable self or emotional labour that is required rather than a detailed knowledge of accountancy procedures, but it is still a requirement.

Waldinger and Lichter provide a detailed analysis of how labour market segmentation is operationalized at the firm level via three mechanisms: ethnic queuing, social network recruitment and access to on-the-job training. At one level, employers may prefer migrants because they have a dual frame of reference, evaluating (poor) pay and conditions against those available not only in the host, but also in the origin country. Furthermore, while wishing to employ workers with the appropriate proficiencies for particular jobs, employers find it difficult to assess the competences and skills of applicants, particularly in respect of attitudes and motivations. They therefore tend to rely either on the social networks of their existing employees, or try to recruit workers with similar characteristics to these employees. In addition, as most training occurs on the job, it depends on relationships with coworkers; therefore, employers have an incentive to appoint individuals who fit in, or who resemble their existing workers. A further theoretical twist to these arguments is provided by the notion of 'interpellation', or the stereotyping processes applied by employers, and the ways in which migrant workers engage with these (Box 2.8).

Box 2.8 Division, segmentation and interpellation in a London hotel

The authors address the question of why migrants tend to be crowded into lower-paid jobs, through a case study of a large hotel situated near Heathrow Airport, in west London. The particular value of this study is that it explores in details the processes of labour market segmentation, arguing that migrants are not just passive victims in this process, but contribute to it. In theoretical terms, the study draws on Burawoy's (1979: 5) notion of interpellation, an Althusserian concept relating to the way that 'employers/managers construct idealized or stereotypical notions of idealized workers'. In the case of a hotel, there is 'dual inter-pellation' as workers have to conform not only to the expectations of managers and co-workers, but also to those of customers.

The starting point of the study is that:

> Managers, employees, and guests all construct their identities and forge social relationships in the interactions and exchanges that take place in a hotel. Questions about gender, nationality, personal style, embodiment, skin color, weight, bodily hygiene, and language abilities (especially when there is an international client base) are crucial parts both of the decision to hire categorically distinctive workers and of the performativity of workplace identities to pro-duce a particular experience of 'hospitality' for the guests.
>
> (McDowell *et al.* 2007: 12)

Given that the delivery of services depends on a range of social com-petences, that are not easily measured in terms of qualifications or technical skills, employers have tended to rely on various forms of stereotyping when recruiting, for example in respect of gender, age, or appearance. Stereotyping in terms of nationality, although interlaced with gender and race, has also become increasingly significant in a labour market such as London's, where migrants dominate employment in the hospitality industry. Migrants were either recruited directly abroad, mainly for management posts, or via employment agencies, in which case the hotel made known its stereotypes to the intermediary.

In this case study, there were particularly strong stereotypes about the deference of Indian workers and the hardworking nature of Polish workers. These assumptions inform the practice of allocating particular nationalities to particular occupations within the hotel, producing a strongly segmented labour force. Each division within the hotel had its own distinctive process of interpellation, with particular nationalities being the focus of these.

There were relatively few signs of resistance to these stereotypes. Instead, migrants tended to conform to them, 'constructing an iden-tity and maintaining a performance that they felt matched their daily tasks and so identified them as appropriate employees for their jobs' (McDowell *et al.* 2007: 21).

Source: after McDowell *et al.* (2007)

The most important consequence of labour market segmentation theory for our purposes is that it is not irrational for an employer to be highly selective in employing workers from particular national or ethnic groups. They may not, of course, be the most suitable, talented or effective potential employees and this is one reason why knowledgeable migrants often cannot utilize their knowledge more fully. Another reason that militates against this is that, in some jobs, employers may put far more onus on features other than knowledge, for example, attitudes. Labour market segmentation is therefore an important theoretical starting point for our later discussions, particularly in Chapters 6 and 7, of the barriers that many migrants encounter in learning and knowledge transactions.

Conclusions: tacit knowledge and migration

We have argued in this chapter that knowledge provides in many ways a more useful concept, or at least a complementary concept, to those emanating from human capital, skills and competences for analyzing some of the economic aspects of international migration. Of particular significance in this debate is the notion of tacit knowledge, and its value to organizers. Tacit knowledge can be disaggregated into different sub-concepts, in particular embrained, embodied, embedded and encultured knowledge (Blackler 2002). There are differences in the extent to which these can be transferred across borders, not least because the latter two are strongly socially situated and based on shared values. While this usually poses obstacles to migrants deploying their other knowledge to the full, it also creates opportunities. Migrants are potentially bearers of unusual forms of knowledge, and those who can overcome very significant obstacles to transacting this can become knowledge brokers. International migration is therefore not only a source of knowledge transfer, but of a complex typology of such transfers. The following chapter will seek to explore the nature of knowledge and knowledge transfers in more detail.

3 Knowledge and knowledge transactions

Mobile knowledge matters

Economists have a longstanding interest in the key determinants of long-term growth. Since the 1980s, new growth theory (developed mainly in the numerous publications of Romer, Grossman and Helpman) has emphasized knowledge as a critical factor, alongside labour and capital. Some authors (for example, Drucker 1993) argue that knowledge is the only meaningful resource and that the capability to create and utilize knowledge is the most important source of a firm's sustainable competitive advantage (Nonaka *et al.* 2000). Lundvall (1992: 8) also made a direct link between economic development and the transfer of tacit knowledge.

Knowledge is distributed across space so we also need to consider its mobility, and of course we are interested in one particular form of mobility, international migration. We start this chapter, therefore, with a specific example, the 'economic miracle' in Mauritius. Africa's story of economic development has largely been one of failure in the last 50 years. Few countries in Africa have been able to sustain high rates of economic growth and catch up with the fast-developing economies in South-East Asia and Latin America. By the early 2000s, only two African countries ranked among the top 50 countries in the world in terms of per capita Gross Domestic Product (calculated on a PPP, or Purchasing Power Parity, basis): South Africa and Mauritius.

At one stage, Mauritius was also a candidate for economic growth failure status. In the 1960s the economy was heavily dependent on production and exports of one commodity (sugarcane). Its population grew rapidly and ethnic tensions deepened. The Mauritian government initially responded with the standard contemporary economic remedies. High tariffs were imposed to promote local import-substituting industries, leading to decreased imbalances in foreign trade, economic diversification and export-based economic growth. These polices, unfortunately, failed to meet expectations.

In the late 1960s, the Mauritian government adopted different policies, prioritizing the attraction of foreign capital via a low tax regime. An Export Processing Zone (EPZ) was established in 1970. The resultant influx in foreign direct investment (FDI) benefited from Mauritius's transnational ethnic networks. Mauritius had a small Chinese community, which had a significant role in attracting FDI from Hong Kong. Textile and clothing-producing firms from Hong Kong moved in and the EPZ quickly became a success. The garment industry accounted for some 80 per cent of investment and 90 per cent of employment in the EPZ. Other manufacturing, financial services and tourism followed subsequently. Mauritius's Indian community also promoted the signing of a double taxation treaty between Mauritius and India, leading to the island becoming an important offshore centre which mediated substantial financial flows to India.

In the period 1973–99, Mauritius experienced an 'economic miracle'. Its total real GDP grew at an annual average rate of 5.9 per cent, and by 3.3 per cent in per capita terms. The equivalent average growth rates in Africa at this time were 2.4 and 0.7 per cent respectively. Economic growth was led mainly by increasing inputs of labour and capital. With decreasing unemployment rates and increasing wages and living standards, Mauritian firms had to focus on higher productivity levels. Total factor productivity generated one-quarter of per capita growth in 1990s, compared to 10 per cent in the previous time period. By 1999 the EPZ accounted for one-quarter of GDP and two-thirds of exports in Mauritius (Subramanian and Roy 2001). Is Mauritius just another success story of a liberal capitalist model, in which Mauritius was able to rely just on its own resources? Romer (1993) thinks differently (see Box 3.1), and argues that international mobility was critical in the transfer of the tacit knowledge that was essential for establishing the garment and financial services sectors in the export zone. In the remainder of this chapter we explore further the nature of knowledge.

What is knowledge?

Knowledge is a fascinating and elusive concept, which has attracted the attention of researchers and philosophers for centuries (Box 3.2) and there is an extensive literature in this area. Yet, it is difficult to find a clear and concise definition of this term. There is a common understanding that knowledge has to do with the emergence and structuring of thoughts, but there is no clear guidance as to which thoughts and which processes enter knowledge creation and transmission.

One major group of definitions refers to philosophy (and logics in particular) and points to particular properties that make thought into knowledge. Knowledge is defined as 'justified true belief' (Dancy 1985). A thought must be *recognized, justified, true* and *believed*. These properties effectively mean that:

- One or more human beings are able to distinguish a particular thought from other ones and recognize it as a separate entity of thought.
- The thought is justified in terms of logic and/or empirical experience, and/or beliefs. This means that beliefs about the thought are based on deductive or inductive reasoning. The justificatory boundary between a

---

Box 3.1 **The importance of mobile ideas: Mauritius's 'economic miracle'**

Romer acknowledges that although the Mauritius population was not rich, they had sufficient surpluses to save to buy simple sewing machines and enter the garment industry. The population was not illiterate, so the government could provide courses on garment production and management. Knowledge of these issues was, after all, a public (non-rival) good, freely available to everybody. Books on the garment industry were widely available and this codified knowledge could be distributed among local entrepreneurs.

What the island lacked, however, was *tacit knowledge* about garment production. Nobody really understood how this industry operated, until entrepreneurs from Hong Kong came to EPZ. This knowledge did not 'leak' from Hong Kong. It was brought in when entrepreneurs were presented with an opportunity to earn a profit on the knowledge they possessed. 'It is quite clear that foreign investors brought to Mauritius a specific set of ideas. Equipment investment (funded by both foreign and domestic firms) followed in response' (Romer 1993: 565).

Romer conceptualized ideas as a special type of public good and hinted that governments should subsidize the use/production of ideas in order to foster economic growth. In Mauritius, foreign entrepreneurs brought an array of ideas in a new line of activity (garment industry), which previously were unavailable in that country. Given their tacit nature, ideas on establishing and operating garment and financial services firms could hardly be transferred in any way other than via the migration of skilled people (entrepreneurs in this case). This was facilitated by the networks of the Chinese and Indian communities living in Mauritius.

Source: after Romer (1993)

---

true belief not being knowledge and being knowledge is, of course, difficult to pin down in some cases.

- Beliefs about the thought accord with facts and correspond to reality, whether for a single person or an entire community or a society. In reality, however, beliefs are states of mind and brain, and subject to rational thought and emotional arousal.

The understanding of knowledge has shifted over time. Since the 1940s, the development of information and communication technologies (ICT) and efforts to create artificial intelligence have had profound impacts on understanding knowledge in relation to information gathering and processing.

Box 3.2 On the origins of knowledge

Socrates and Confucius considered knowledge to be both a virtue and a resource, which could be developed through learning. Both philosophers pointed to self-knowledge as being highly significant. This included knowing what one knows and knowing what one does not know. This higher level of recognition enables an individual to use his or her knowledge effectively and avoid mistakes in the areas of ignorance. While both philosophers admitted that knowledge could be obtained via inspiration (including divine ones), Socrates emphasized that knowledge also included the ability to defend one's ideas through reasoning or by pragmatic results.

Plato, in his dialogue *Meno*, presents ideas as eternal entities, which are independent from human beings. There could be no new knowledge and all human thinking is just a recollection of 'eternal ideas'. Intelligence, according to Plato, is a process of transforming experience into knowledge. Plato's eternal ideas were later rejected by Aristotle and most other philosophers.

Since the seventeenth century there has been strong development of epistemology, concerned with the theories of the sources, nature and limits of knowledge. While some philosophers tried to marry Plato's opinions on innate ideas with acceptance of the values of data and ideas derived from experience (e.g. Pascal, Descartes, Spinoza), most empiricists (for example, Locke, Hume and Mill) discarded the notion of innate ideas. John Locke, for example, considered knowledge to be the outcome of mental activities. He recognized two kinds of so-called 'proper knowledge' – intuitive and demonstrative. Below the rank of 'knowledge proper', there is a third degree of knowledge, not strictly entitled to the name. This is our sensitive apprehension of external things, or of real objects other than ourselves. Locke thought that 'whatever comes short of proper knowledge, with what assurance soever embraced, is but faith or opinion, but not knowledge, at least in all general truths' (Locke 1975, *Essays*, IV, iii: 14). For further reading on epistemology, see Dancy (1985).

Source: after various sources within http://en.wikipedia.org/wiki/Knowledge

---

Many authors argued that the fluid and organized character of information about certain topics are basic properties of knowledge. Information-based definitions of knowledge argue that knowledge is gained and preserved by instruction, acquaintance, enlightenment and/or learning. Another perspective is that different kinds of knowledge refer to diverse ways of obtaining knowledge. Some knowledge is established via experience and converted into practical skill, as knowledge of life. The experience may have the character of trial and error, or imitation. The more general kinds of knowledge are usually acquired as facts, truths and principles. These rely on storage and transmission of information, and are learned throughout time. Finally, creative minds are able to derive new knowledge from facts and principles via a process of enlightenment. The complexity of these structures translates into different types of knowledge, which are organized into hierarchies. Descriptive knowledge, for example is about objects, facts and relations between them. In contrast, procedural knowledge is how to use the descriptive knowledge, and meta-knowledge is information about knowledge.

Analogies with ICT are rather mechanistic, as they ignore the cognitive character of knowledge. Information first has to reach the human mind in order to become a part of knowledge. Clear perception of facts or truths, apprehensions, and familiar cognizance and awareness of principles all help to form knowledge. Another perspective emphasizes the interaction of cognitive processes with the person's mental environment, which includes beliefs and emotional states. Knowledge, therefore, possesses both rational and emotional components of mental activities (Box 3.3). This view of knowledge creation opposes the notion that data is a fundamental component of information, and that information is a fundamental component of knowledge. Instead, data emerge only after we have information, and information emerges only after we already have knowledge. Yang (2003) also considered that knowledge is understanding about reality through mental correspondence, personal experience, and emotional affection with outside objects and situations. Hence, there are three inter-related facets of knowledge – explicit, implicit, and emancipatory (or emotional affection).

It is clear therefore that knowledge can be assessed from diverse points of view. Turning to the identification of different types of knowledge, the starting point for most social scientists, of course, is Polanyi's (1958; 1966) seminal work that distinguishes between explicit and tacit knowledge. Explicit knowledge is easy to articulate, codify and transfer in different formats, since it is formal and systematic. Tacit knowledge, on the other hand denotes all intellectual or corporeal capabilities and skills that the individual cannot fully articulate, represent or codify. Tacit knowledge is developed from direct experience, observation or interaction in which one largely learns by doing and understanding is often at a subconscious level.

The concept of tacit knowledge is, however, problematic. Styhre (2004: 178) notes that tacit knowledge is 'white spots on the map in between the well-known areas of explicit knowledge' and considers that the concept is little more than an umbrella term for knowledge which cannot be represented, arguing that 'rather than distinguishing knowledge into a tacit and an explicit component, knowledge arises from two distinct individual capacities, intuition and intellect'. In some cases analytical thinking dominates intuition and, conversely, there are other forms of knowledge where intuition dominates the intellect and relies on synthetic thinking. The human mind

---

Box 3.3 Sense and sensitivity: is rational thought sufficient for know-ledge transfer?

Antonio Damasio, a neuroscientist, challenges Descartes' assumption that when we are being logical, we are excluding emotions. Damasio argues that emotions are essential to rational decision-making processes. He emphasizes the importance of emotion in decision making by interacting with patients whose emotional centres in their brains have been damaged.

In his book, *Descartes' Error* (1995: 193), Damasio writes:

> You had a person who had been doing very well in his or her life – someone who had relationships, friendships, marriage, and a successful career. And then because of a stroke or a tumour, every-thing changed. And the change took place in the realm of day-to-day decision making, not in the realm of knowledge and skills. They could speak perfectly well. They could deal with the logic of a problem. They could learn new things. But, all these people shared one common trait: their emotions were compromised. . . . Even though they scored in a normal range on all standard measures of intelligence, somehow they couldn't navigate the branching deci-sion trees of everyday life.

Damasio's book is relevant for wider research on behavioural economics and neuroeconomics. The transfer of knowledge may be intended as a completely rational process, but is always shaped by specific social and psychological conditions. Rational thoughts alone are not enough for knowledge transfer. As demonstrated by Kahneman (2002), for example, emotions, intuitions, thoughts and preferences are at least as important for human decision making as rational behaviour. They are present in most human interactions, and no study of knowledge trans-fer between individuals and/or societies can abstract from them.

Source: after Damasio (1995)

---

can make sense of very complex realities, even when not all the components and their mutual relations are fully understood. However, for Styhre, knowledge is a fluid and elusive capacity. The importance of intuition in know-ledge is also emphasized in a quotation from Einstein:

> I sometimes *feel* I am right, but do not *know* it. When two expeditions of scientists went to test my theory I was convinced they would confirm my theory. I wasn't surprised when the results confirmed my intuition, but I would have been surprised had I been wrong. I'm enough of an artist

to draw freely on my imagination, which I think is more important than knowledge. Knowledge is limited. Imagination encircles the world.

(Quoted in Taylor 2002: 12)

Despite these theoretical critiques, tacit knowledge is an important concept for our purposes. Because the holders of tacit knowledge do not completely understand what exactly they are doing and/or are unable to successfully express this, it means that tacit knowledge sticks to the individual and is difficult to transfer other than via personal contact. Tacit knowledge usually is shared through highly interactive conversation, storytelling, observation and/or some form of shared experience. Here we note three different attempts to disaggregate the notion of tacit knowledge.

First, Boisot (1998: 57) argues that there are at least three very different variants of tacit knowledge: (1) 'Things that are not said because everybody understands them and takes them for granted'; (2) 'Things that are not said because nobody fully understands them. They remain elusive and inarticulate'; and (3) 'Things that are not said because while some people can understand them, they cannot costlessly articulate them'. The first variant refers to common knowledge such as 'water is wet', although when it comes to defining a notion such as 'wetness' this is inevitably complex. The second kind of knowledge refers to certain embodied talents and skills, for example in the field of arts, but also in management or science. The first and second variants of knowledge are acquired via personal experience, intuition and/or enlightenment and are very difficult to transfer, if at all. The third variant refers to knowledge transferable via shared experience, either via in-group exchange or mobility between groups of individuals and/or organizations – although at a cost, so that this is not universal practice.

Second, Lundvall and Johnson (1994) refer to four types of knowledge:

- know-what: a broad knowledge about facts which is very similar to information;
- know-why: an understanding of scientific principles;
- know-how: specific skills ranging from artisan aptitudes to the ability of business people to assess market opportunities; and
- know-who: the density and strength of social networks.

The first two types of knowledge broadly equate to Polanyi's definition of explicit knowledge. The last two are more tacit in character, but not exclusively so, as some types of know-how can be codified and transferred (for example, via patents or copyrights).

Tacit and explicit knowledge should not be seen as separate entities, but as interlinked. Tacit and explicit knowledge can be combined and mutually, if not fully, converted. Nonaka and Takeuchi (1995) identified four types of knowledge transfer, involving different combinations of tacit and explicit knowledge:

- socialization (from tacit to tacit);
- externalization (tacit to explicit);
- internalization (explicit to tacit); and
- combination (explicit to explicit).

Conversion of tacit to other forms of knowledge requires an intimate personal experience and good intuitive understanding of the issues concerned (see Box 3.4 for an example of externalization).

Box 3.4 Converting tacit to explicit knowledge: Ibn Khaldun

Ibn Khaldun (1332–1406), born in Tunis in 1332, was a politician, scholar and diplomat in Tunis, Fez, Granada, Tlemcen, Alexandria, Damascus and Cairo. Khaldun's (1969, translated and republished) most famous book is *Muqaddimah*, which draws on these experiences. In essence, this is a grand sociological work, which includes a sociology of knowledge. This includes the concept known in economics as the Khaldun–Laffer Curve (the relationship between tax rates and tax revenue follows an inverted U shape).

> It should be known that at the beginning of a dynasty, taxation yields a large revenue from small assessments. At the end of a dynasty, taxation yields a small revenue from large assessments. . . . When tax assessments and imposts upon the subjects are low, the latter have the energy and desire to do things. Cultural enterprises grow and increase, because the low taxes bring satisfaction. When cultural enterprises grow, the number of individual imposts and assessments mounts. In consequence, the tax revenue, which is the sum total (of the individual assessments), increases. . . . When the dynasty continues in power . . . custom and needs become more varied because of the prosperity and luxury in which they are immersed. As a result the individual imposts and assessments upon the subjects, agricultural labourers, farmers and all the taxpayers, increase. Every individual impost and assessment is greatly increased, in order to obtain a higher tax revenue. . . . Eventually the tax will weigh heavily upon the subjects and over-burden them. Heavy taxes became an obligation and tradition, because the increases took place gradually, and no one knows specifically who increased them or levied them. . . . The assessment increases beyond the limits of equity. The result is that the interest of the subjects in cultural enterprises disappears, since when they compare expenditures and taxes with their income and gain and see the little profit they make, they lose all hope. Therefore, many refrain from all cultural activity. The result is that the total tax revenue goes down, as individual assessments go down.
>
> (Khaldun 1969: 230–1)

This and his other books represent a supreme example of the conversion of tacit knowledge into codified knowledge. At various times, Khaldun had been a prime minister, chancellor and supreme judge to several sultans. During his long career, he observed (and sometimes assisted in) the rise and fall of empires in Northern Africa and developed considerable personal knowledge of the social, cultural and economical determinants of the unfolding history. The tacit knowledge he acquired was converted into explicit knowledge in his writings. His magnum opus, *Muqaddimah*, is infused with the central concept of 'asabiyah', that is social cohesion. This cohesion arises spontaneously in tribes and other small kinship groups; and can be intensified and enlarged by a religious ideology. Ibn Khaldun's analysis examines how this cohesion carries groups to power but contains within itself the seeds – psychological, sociological, economic, political – of the group's downfall, to be replaced by a new group, dynasty or empire bound by a stronger (or at least younger and more vigorous) cohesion.

Source: after Rabi (1967)

Third, Blackler (1995) argues that knowledge is not constituted of discrete packages of information, which can be separated from the environment in which it was formed. Instead it is:

- influenced by technology and organizational systems;
- is 'situated' in relation to social, cultural and physical environments;
- constantly changing and evolving;
- pragmatic, in that individuals only learn what they feel it is necessary to learn;
- open to query, even though characterized by hegemony and hierarchy.

For the purposes of this volume, Blackler's (2002) review of his own earlier work on the nature of knowledge, and those of Zuboff (1988), Berger and Luckmann (1966), and Brown and Duguid (1991) are particularly useful. He identified four main types of tacit knowledge – embrained, embodied, encultured and embedded – as well as encoded knowledge. We have already defined these and started to discuss their value in context of migration studies in Chapter 2.

Blackler's approach is particularly useful in examining knowledge transfer via international migration. Embrained and embodied knowledge are necessarily indivisible from the individual, and so are fully transferable via human mobility. Encultured and embedded knowledge are more grounded in the relationships between individuals, in particular settings; that is, they represent specific forms of relational knowledge. In so far as these settings are not transferable or replicable (but international franchising represents an attempt to achieve this end), then they are – at best – only partly transferable through human mobility. A more negative reading would emphasize that they are necessarily disrupted by human mobility, and so are non-transferable. However, this would be to deny individuals the capacity for reflexivity and comparison between places and institutions. Hence, migrants do take encultured and embedded knowledge with them, even if, in themselves, these are difficult to transfer to others who lack shared values and knowledge.

This is not, however, to deny that there are obstacles to applying all types of knowledge in different settings. For example, transferring embedded knowledge between organizations is problematic because it is highly context-specific and resides in an organization's inter-related systems of physical, human and organizational relationships (Empson 2001); this is discussed further in Chapter 6. Finally, while it is important to distinguish between different types of knowledge, one of the keys to utilizing and valorizing knowledge is how these are combined (see Box 3.5). In this sense, all forms of knowledge are relational, and none are transferable without transforming their potential economic value.

***Box 3.5* Technology and tacit knowledge transfer: Canadian investment in China**

Technology transfer from developed to developing countries is a major channel for the international circulation of knowledge. What types of knowledge are transferred, which mechanisms are used and what are the characteristics of the learning process in recipient firms? Marcotte and Niosi (2000) explored these issues in a study of a sample of 28 Canadian firms transferring technology to China. The companies were from the high- and medium-tech manufacturing industries (electronic, telecommunication and electrical equipment; transportation equipment; industrial machinery; metal products and chemicals) and were interviewed during the period June 1995–May 1997.

Both explicit and tacit knowledge were transferred. Blueprints and manuals and technical assistance were important transfer mechanisms (Figure 3.1). However, the transfer of tacit knowledge (identified with know-how and technical services) was more important than explicit knowledge (trademarks and patents). The importance of tacit knowledge transfer was highlighted by the emphasis placed on training Chinese staff in the recipient firms. Most difficulties in transferring knowledge were related to cultural and institutional contexts (bureaucracy, language and cultural differences).

Marcotte and Niosi hypothesized that learning and technology transfer evolve over three distinctive phases. In Phase 1 (learning-by-doing) the recipient firm learns to solve the immediate problems of technology acquisition and adaptation. Phase 2 (learning-by-adaptation) involves deeper insights into the principles underlying the technological and organizational changes necessary for the integration of technology. In Phase 3 (learning-by-creating), large organizational innovation

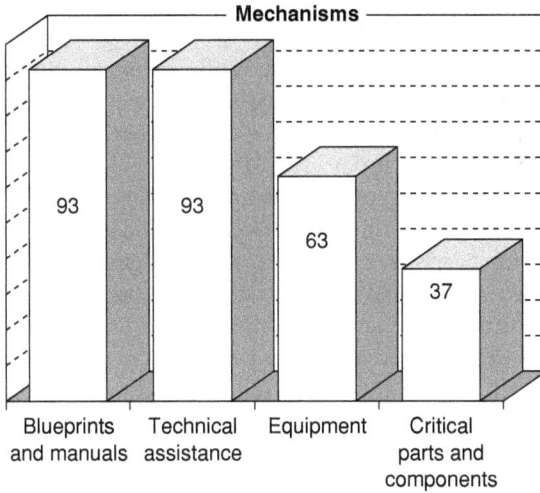

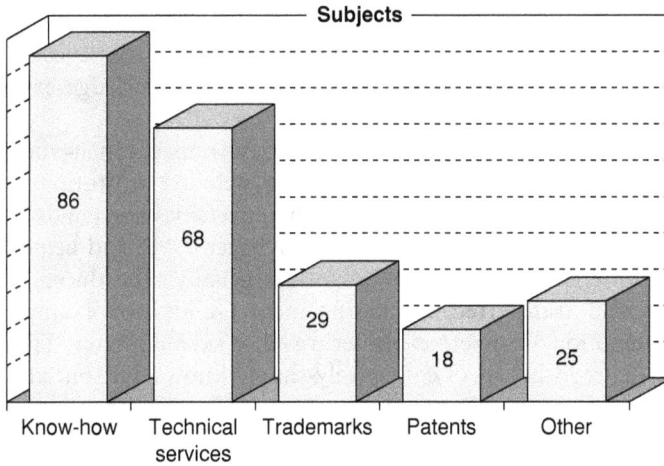

Figure 3.1 Importance of mechanisms and subjects of transfer agreements in
technology transfer (%)

Source: after Marcotte and Niosi (2000)

associated with new products and processes follow. Phases 2 and 3 are
characterized by higher levels of tacit knowledge than Phase 1. Marcotte
and Niosi assumed that the capacity to adapt technology (Phase 1) would
take less time than the capacity to design new goods from the same basic
technology (Phase 3), and found some evidence for these assumptions.
While some 17 out of 28 Chinese firms (60 per cent) acquired the capacity to adapt the technology, just eight companies out of 28 (28 per cent)
evolved a capacity to design new goods from the same technology.

Source: after Marcotte and Niosi (2000)

Tacit versus explicit knowledge transfer: historical perspective

Before systems for codifying knowledge were invented, all knowledge transfer happened face-to-face. Knowledge sharing was a purely personal experience, strongly dependent on trust. Of course, early societies lived in relatively tightly-bounded territories. Transfer of knowledge over long distances was difficult due to high travel costs and differences in local cultures. However, the invention of writing enabled the codification of knowledge, facilitating its transfer over space and time. Knowledge diffusion, however, remained limited even after formal writing systems had been established. Early writing systems were bound to specific countries and/or cultures, were difficult to master and were accessible only to privileged segments of the population who could read. Most knowledge exchange still relied on face-to-face contacts. It was not until the Hellenistic period that the first signs of a quasi global culture, business and travel emerged, at least in Europe and the Middle East. With the development of the Greek and Roman alphabet, and later printing techniques, the transfer of explicit knowledge was made cheaper, easier and increased in importance. Tacit knowledge, however, remained expensive to transfer over long distances, because the major bottlenecks – the costs of travel and socio-cultural differences between knowledge exporting and importing countries – persisted.

The growing demand for new technology transfers in medieval Europe gave rise to new means of transfer. These included institutions with cross-national membership, such as the Church, universities and guilds. They essentially acted as communities of practice (Wenger 1998) and helped to bridge cultural differences. Tacit knowledge was usually reproduced via apprenticeships and transferred via journeymen. Guilds, for example, enabled seasonal and long-term cross-border travel of skilled labour. They operated not only as repositories of collectively-shared knowledge, but also provided apprentices and travelling journeymen with accommodation, work and vocational training in exchange for below-market wages. Apprentices and journeymen acquired knowledge via experiential and observational learning. The accumulation of some types of knowledge required years and was only available in specific places. This means of knowledge transfer could require long-term and long-distance migration and was, of course, costly and slow. It is no accident that most technological advance was sited in urban areas in the Low Countries and Northern Italy, and later England and Central Scotland, which were densely populated and enabled reduced information and travel costs (Epstein 2005).

Printing made the diffusion of explicit knowledge cheaper and faster, but the costs of tacit knowledge transfer remained substantial until growing urbanization, integration of financial markets and advances in transport technologies led to cheaper and easier travel. Since the nineteenth century, revolutionary changes in transport and communication technologies have made massive international transfers of people, money, goods and knowledge possible at hitherto unimaginable levels of prices, speed and accuracy. On this basis, it can be argued that new transport and communication technologies support dissemination of explicit and codified knowledge, while tacit knowledge transfer has decreased in importance for several reasons:

(1) Development of the knowledge-based economy is fuelled by scientific knowledge, which is easier to codify and transfer than other forms of tacit knowledge. Firms and other organizations can rely more on patents and other forms of property rights to capture scientific knowledge rather than on human mobility.
(2) Modern ICT technologies support delocalization of knowledge, but simultaneously enable the interaction of individuals with people in distant places (for example, via the internet and/or videoconferences). This means people can develop more social capital and trust via virtual networks than via personal presence in some instances.
(3) The modern world is globally integrated and there has been a remarkable homogenization of languages, ideas, technology, products, cultures and organizational forms. Globalization erases differences between places and decreases need for tacit knowledge transfer, as these places are becoming increasingly similar.

Given our emphasis in this volume on the neglected role of international migration, we do of course contest these arguments.

First, the boundaries of firms and other organizations are uncertain and porous, and human mobility plays a key role in spreading knowledge, whether via communities of practice, individual knowledge brokers or unplanned overspills, for example through labour market turnover. Some organizations (firms, universities, public governance bodies) try to eliminate, or at least reduce, such leakages via control of intellectual property rights (patents, designs), forcing employees to sign confidentiality agreements, or banning them from employment with competitors for certain periods after leaving their present employer. These controls usually have limited effectiveness.

Patents, designs and other forms of codified knowledge cannot capture experience transformed into tacit knowledge. And it is difficult to effect legislation or regulation to prevent the unplanned transfer of tacit knowledge. It could be argued that tacit knowledge (transferred via face-to-face interactions) is only important in incremental innovations, while radical innovations arise either from moments of enlightenment in gifted individuals or from the application of explicit knowledge codified in science and technology methods. However, modern science is based on teamwork. Work in universities and research laboratories relies as much on formal training as on mutual engagement among participants in communities of practice. Learning via observation and analogy is often at least as important as didactic learning or learning via experience. When producing a novelty, it is impossible to separate parts of the knowledge chain, such as routines, conversations, meetings, scripts, memory, stories, and other soft technologies, grafted onto the technologies of formal knowledge acquisition and application (Amin and Cohendet 2004: 108).

Second, it has to be acknowledged that the development of formal writing systems, and later ICT, has led to more codified and less personal knowledge transfer. However, this poses the question of whether transmissions of codified knowledge, and email and online communications, actually increase or decrease levels of trust and social capital. The effects of trust are less visible perhaps when knowledge is transmitted via electronic media, but it does not mean there are no requirements for trust (and face-to-face contracts). Any recipient of knowledge must make an internal decision as to whether the transmitted knowledge is trustworthy and useful, or not.

Third, globalization is often held to diminish differences between social settings and places, due to – among other factors – easier and cheaper human mobility. Global growth in human mobility, however, need not necessarily result in a 'global village' and the elimination of geography. Several factors contradict the supposed hegemony and reach of globalization.

- Globalization seems to result in greater homogenization of socio-cultural milieus in consumption than in production. Different national production systems are embedded in diverse and durable institutional frameworks that endure despite growth in international trade. Institutional rigidities tend to restrain convergence. This is exemplified by Japanese carmakers' attempts to transfer lean production methods to the United States. Lean production is, among other things, based on long-term employment and lifelong on-site learning. Employees are offered stable jobs in exchange for adjustments of wages, tied to company profitability. These production principles conflict with how labour markets operate in the USA. Job protection is low, workers are hired for shorter time periods and receive less specific on-site training. Globalization is unlikely to eradicate the significance of such national and/or supra-national boundaries and institutions. In fact, the importance of tacit knowledge transfers via labour mobility has increased with the shift to more flexible specialization and the introduction of post-Fordist modes of production and governance. Specialization has supported competitiveness and innovations, as flagship companies could select suppliers from a large pool of medium-sized companies. The resulting horizontal and vertical cooperation within clusters required substantial inter-firm communication. Competing second- and third-tier suppliers learned from one another and adopted best-practice technologies and organizational routines. With multinationals expanding their operations worldwide, the scope for knowledge creation and transfer also became global, but regions still remain important areas of innovation and diversity (see Chapter 5).

- Innovation promotes divergence. The greater the inputs of knowledge, the higher the rates of innovation and the higher the divergence. Recent decades have seen massive increases in international trade in goods and services, with implications for divergence.
- Production occurs in specific times and places. Inputs of knowledge are essentially 'person-embodied' and subject to human mobility. However, while the volume of international travel has increased enormously in recent decades, globalization has resulted more in capital than human resource flows, not least because human mobility is constrained more by regulation. National states continue to be the key sites in the regulation for international mobility, as argued in Chapter 4.

In summary, despite the growth of increasingly efficient ways of codifying knowledge, and of transferring such knowledge over space and time, much tacit knowledge remains personal knowledge, and often requires face-to-face contacts for learning and efficient knowledge transfer. Moreover, it has been argued that with the increasing ubiquitification of explicit knowledge, the importance of tacit knowledge transfers has become even more significant in competitiveness (Maskell and Malmberg 1999) – and, therefore, the importance of human mobility in knowledge transfer.

The economics of tacit knowledge transfer

Solow's two seminal papers on economic growth (1956; 1957) argued that technological progress might allow the economy to expand output without necessarily adding more labour and capital. This model paved the way to the concept of total factor productivity, but did not explain the origin of technological progress. Invention, innovation and ingenuity were all considered exogenous factors of growth. Subsequently, Romer (1990) endogenized the rate of technological progress. Endogenous growth models consider knowledge to be a public, non-excludable and non-rival good. Some kinds of knowledge may incur high costs, but once the knowledge has been created, its use by one individual or organization does not prevent another individual or organization from using it, unless patents or copyrights protect the knowledge. Moreover, the use of knowledge by one user also does not prevent simultaneous implementation by other users.

Endogenous growth models suggest the essential importance of the 'production' of new technologies and human capital. Individuals and businesses have incentives to innovate in order to exploit an advantage over their competitors, thereby improving their own productivity. However, over time some of the knowledge associated with the innovation 'spills over' to other economic actors, which increases those actors' ability to innovate. Whereas knowledge may be costly to produce, it is cheap, and often almost costless, to reproduce. Thus, according to endogenous growth theories, the total cost of a design does not change much, whether it is used by one person or by one million.

Although knowledge *is* a public good with property rights that are rarely enforceable (except for patents and designs), this does not mean that transfer is costless. Romer's (1990) assumptions about costless transfer apply to explicit knowledge goods, such as databases transferred via electronic media. But even this type of knowledge transfer is not completely costless, as its effective use requires the time of, and learning by, knowledge recipients. Learning is not costless and the expense is *at least* equal to the cost of recipients' time.

The costs of tacit knowledge transfer are even more substantial. Putting aside transfers via electronic means – for example by videoconferencing – tacit knowledge is bound to its human bearer, and the costs of transferring tacit knowledge overlap with costs of human mobility. In general, the costs of tacit knowledge transfer (C_{TKt}) be expressed as follows:

$$C_{TKt} = TRc + Oct + FITc$$

where

- *TRc* is costs of travel and reallocation for individual migrants;
- *Oct* is the opportunity costs of migrant time; and
- *FITc* is the costs of formal and informal training and socialization in the place of migration.

The opportunity costs of a migrant's time are usually associated with wages, but may also include the psychological costs of migration, linked to disruption of previous social networks and the cultural environment (see Chapter 2 on human capital theories). The costs of training include both the costs of formal training, such as school or university fees, and the costs of informal training (socialization and adaptation costs). The opportunity costs of time are likely to be the major part of the total costs for most migrants. The costs of travel and re-allocation, on the other hand, may be substantial in the case of unregistered migrants paying fees to traffickers, while the costs of formal training are likely to impact on students in particular. The costs of socializing can be relevant for some migrant groups, such as business experts and executives who use golf clubs, dinners and parties for such purposes.

The benefits of knowledge transfer can be measured both at the micro and macro levels. At the micro level, there are significant benefits in terms of the increased productivity of a firm adopting specific knowledge. At the macro level, the benefits are visible in changes in the total factor productivity of the economy, but with different consequences for countries of migration origin and destination (see the discussion of the aggregate welfare effects of migration in Chapter 2). Of course, remittances, enhanced human capital and returned migration can shift this zero-sum game to a win–win outcome depending on the knowledge transactions which are effected by migration.

The discussion of human capital theories (Chapter 2) has already touched on how psychological factors, and social, cultural and economic institutions, shape migration. Here we summarize their impact on the efficiency of knowledge transfer. The efficiency of knowledge transfer via migration can be formalized into a function:

$$E_{TKt} = f\{PCM, SC, SCED, OC\} + error$$

where

- E_{TKt}, the dependent variable, can be expressed in terms of the speed and density of diffusion of a particular type of knowledge, such as the introduction of new surgery techniques, or the implementation of quality management systems;
- *PCM* is the personal capabilities of a migrant and is a vector of variables including education and training, level of understanding of the knowledge issues, language ability, communication and presentation skills, and motivations for knowledge transfer;
- *SC* stands for social capital in the place where the knowledge transfer is realized, and is a vector of variables including trust and shared/or super-ordinate social identity;
- *SCED* stands for socio-cultural and economic differences and is a vector of variables including shared norms and beliefs;
- *OC* stands for organizational culture and is a vector of variables related to how conductive an environment is for knowledge adoption and sharing; and
- *Error* stands for knowledge transfer 'noise' generated via misconceptions and/or faulty information.

Some of the above-mentioned independent variables can be formalized directly and externally (for example, level of education or training, quality of management systems), while others are assessed via indirect and self-rating techniques (such as, motivations or social skills). It is also possible to construct proxies for language and socio-cultural differences, trust and some of these other factors, although the reliance on such quantitative, and often surrogate, measures can be problematic.

Relations between the dependent and independent variables in such a function can be very complex. Consider a hospital in country X, which has developed new methods of organ transplant. Migrant doctors from countries A, B, C, D, E and F were present in the hospital at the time when the new method was developed and were offered an opportunity to learn and transfer it to their home countries. Countries A, B, C and D have very similar socio-cultural and economic settings to country X and there is considerable potential to transfer this knowledge. The migrant doctor from country B, however, has a lower degree of social capital and is less able to communicate and present his new knowledge to colleagues than the doctor from country A. Knowledge transfer from country X to country B is therefore likely to be slower than to country A. The enthusiastic and motivated migrant from country C returns to his home hospital, but the hospital management decides not to support the introduction of new organ transplant methods and – in extremis – the migrant finds his or her job has been abolished. The migrant doctor from country D is also enthusiastic and motivated but is rather young and his or her colleagues both in the host and home countries do not fully accept his experience. Country E is comparatively rich, but has different cultural norms and does not allow for organ transplants. Country F does allow organ transplants but is too poor to provide the basic infrastructure for this medical intervention. In these circumstances, doctors from countries C, E and F have little motivation to try and transfer knowledge, and even less prospect of being successful (see also Box 2.7 for an empirical illustration of medical knowledge transfer).

Modes of knowledge transfer and learning

Individuals and organizations can acquire, accumulate and transfer knowledge in a number of ways, all of which involve some form of learning. Learning is not limited to attainment of formal (explicit) knowledge, but also relates to the acquisition of values, skills and attitudes. Learning can also result in the formulation of new mental constructs and generate long-term behavioural changes. There are several ways in which an individual or organization can learn (Kolb 1984), although the relative importance of these is contingent.

- The most basic way of learning, *observational learning*, is known as *imitation* or modelling. The learner observes the conduct of 'model' (or benchmark) personalities or organizations and performs an act of repetitive behaviour in learning from them. Imitation enables the transfer of both simple knowledge and routine skills, and also some complex social behaviour such as how to support other group members. Learners usually do not confine themselves to simple repetition of observed activities, but actively process the knowledge obtained from their observations.
- *Learning via analogies* relies on recognition of structural features and finding a common denominator on two and more issues. This kind of learning particularly facilitates creativity and flexibility.
- *Learning via abstract conceptualization* is based on acquiring knowledge on general principles that are valid for a range of issues. This method enables rapid and frugal acquisition of knowledge on a potentially extensive number of issues and is particularly suited for transferring

scientific knowledge. However, learners are often unable to appreciate and extend principles to novel situations, unless there is a close relation between the abstract principle and a specific situation.

- *Learning via concrete experience* refers to active experimentation and the self-correcting role of feedback. People and/or organizations are presented with initial information about the structure of a task, which provides a basis for initial judgements and solutions. The outcomes of these initial judgements and calculations produce two feedback mechanisms: a direct effect on subsequent judgments and actions, and an indirect effect based upon a person's interpretation or evaluation of the outcomes.

There is also an extensive body of writing on action learning, which is grounded in theories of learning from experience, as practised collaboratively with others through some form of action. Marsick and O'Neil (1999: 162–3) identify three types of action learning: scientific, experiential and critical reflection. Scientific is similar to action research, but biased towards learning. In experiential learning, action, reflection, theory and practice are combined. Finally, in the critical reflection school of learning, individual learn from reflection on the assumptions and beliefs that mediate practice.

Most learning is a hybrid, combining two or more types. The learning methods used are determined by both the nature of the learning task and the capabilities and preferences of the learner. Particular tasks also require different levels of knowledge codification and transfer. For example, abstract conceptualization is usually easier to codify and transfer, but may be sub-optimal when a significant part of the knowledge content is tacit. Observational learning is probably the most common form, and one of the more efficient methods of knowledge transfer where most of the knowledge content is tacit.

There is some experimental evidence of the appropriateness and efficiency of particular learning techniques for particular knowledge transfer tasks. Nadler *et al.* (2003), for example, used student groups to simulate job contract negotiations between an employee and employer, and commercial real estate negotiations between a development company and a city planner's office. Four groups of students learned negotiation techniques in the four different ways of learning highlighted in the bullet list above. Students trained via observational and analogical learning were more likely to construct effective trade-offs among issues than students in the control group and who had been trained in the abstract conceptualization and concrete experience modes. Interestingly, students in the more effective groups found it difficult if not impossible to articulate the learning principles that had helped them to improve. Nadler *et al.* assume that they acquired tacit knowledge, which they were unaware of having, and unable to articulate in their written work. In particular, observational learning, or learning via imitation, proved most efficient when dealing with the tacit knowledge of negotiation but, at the same time, was most elusive to express and codify. Instead, as expected, such knowledge was most easily acquired via face-to-face networking.

Most learning methods, whether by experience, observation, and/or inference, involve interactions with other individuals, either directly or indirectly (via absorption of codified knowledge) and are therefore mediated by social networks. This also applies to organizations. Argote and Ingram (2000), for example, consider that knowledge transfer is a process by which one unit of an organization (a group or department) is affected by the experience of another. The efficiency of the transfer inevitably depends on the intensity and nature of the interactions within and between these units. This emphasizes the role of social capital. Lin (2001: 56) argues that social capital is 'resources embedded in one's network or associations'. This can mediate transactions cost, as Fukuyama (2001) explains: 'The existence of a certain (i.e. specific) set of informal values or norms shared among members of a group that permit cooperation among them and in economic spheres reduces transaction costs'.

Social capital relates to an individual's ability to make weak and strong ties, and to trust others, within a system. Trust makes social life more predictable. It creates a sense of community, making it easier for people to work together. Given that it increases predictability, social capital can be important for decreasing transaction costs in the movement of labour and knowledge.

Turning from learning to knowledge transfer, we start by noting that differences in types and levels of knowledge between two or more persons/ organizations are a necessary rather than a sufficient condition for knowledge transfer to be realized. There also has to be a 'sharing context' between the parties. This means that, whether the knowledge transfers are between or within organizations, they are inevitably shaped by institutional arrangements and social capital. The amount and quality of social capital (in terms of the number and strength of social ties) determines the speed and quality of knowledge transfer.

All organizations are 'institutional' in nature, and are embedded in networks of institutions. These are extremely diverse for each specific society, time-period and place. This diversity, in turns, influences the development trajectories, management practices and organizational routines of firms. The institutional environments also mediate the selection of knowledge to be adopted, implemented or transferred. Successful knowledge transfer, to a considerable degree, depends on bridging differences in institutional settings. To be effective, managers in an importing organization must be familiar with both the organizational and the socio-cultural context in which the knowledge originated. For example, technology transfer carried out without understanding its interaction with the social and environment, and the existence of different regulatory mechanisms, can lead to malfunction, or even tragic disasters, as illustrated by the case of Union Carbide's Bhopal plant (Munir 2002).

Table 3.1 Types of knowledge transactions by type of migration

Type of migrants Time perspective	*Émigrés*	*Returnees*
Short-term	*Knowledge acquisition* Integration by émigré to a host society. Building trust and social capital, creating social identity. Imitation, learning by doing.	*Knowledge transfer* Conceptualization of acquired knowledge, transfer of both tacit and explicit knowledge via didactical learning and demonstration.
Long-term	*Knowledge transfer* Conceptualization and ordering of knowledge. Drawing principles, converting some tacit knowledge to explicit knowledge.	*Knowledge acquisition* Reintegration to home community, rebuilding trust. Acquiring knowledge of recent developments in home society. Drawing principles, and abstract conceptualization. Potential conversion of some tacit knowledge to explicit knowledge.

Source: authors

The more similar the social and cultural profiles of two organizational populations, the easier it is to transfer knowledge between them. Similarly, a host population can more easily identify with an immigrant who has a similar social identity than with a migrant coming from a very different cultural and/or social environment. The latter migrant will usually experience more problems in transferring his/her knowledge to the host community but, in turn, such migrants have advantages in transferring new knowledge back to their country of origin, where their social capital is stronger (Table 3.1).

A similar identity between organizational populations can be at least as, or more, important for knowledge transfer via individuals moving between them than the quality of the knowledge itself. Experimental research by Kane *et al.* (2005), for example, found that people evaluate in-group members as more trustworthy, honest, loyal, cooperative and valuable to the group than out-group members (Box 3.6). Individuals may feel more comfortable sharing knowledge with groups with whom they share a social identity than with groups with whom they do not share such an identity. Hence, building trust is an important precondition for deciding whom to learn from or to share knowledge with. If a host society is radically different from the society of origin, building trust may take many years and, in the face of persistent discrimination, may be impossible (see Box 3.7).

***Box 3.6* Transfer of tacit knowledge via personal mobility: Kane's experiment**

Kane *et al.* used experimental methods to evaluate the importance of social identity and knowledge quality in knowledge transfer: 144 students performed the task of constructing origami sailboats (the Japanese art of paper-folding). Twenty-four groups of students were formed, with six participants in each group. The students were induced to think of themselves as either a six-person group comprised of two three-person work groups (super-ordinate identity condition) or as two separate three-person work groups (no super-ordinate identity condition).

Half of the work groups were trained in a slightly superior production routine, which was more difficult to learn, but enabled them to reduce the number of folds from 12 to 7. Students, however, were not told that different work groups were trained to use diverse production routines – superior and inferior ones. Midway through the task, a member from a different group transferred into each group. Rotating members possessed either inferior or superior knowledge on origami production, and the resulting knowledge transfers were observed:

- as expected, knowledge was more likely to transfer from a rotating member to a recipient group when the rotating member possessed a superior rather than an inferior knowledge;
- knowledge was easier to transfer by a rotating member to a recipient group when the mover and the group shared a super-ordinate social identity;
- when the rotating member did not share a super-ordinate identity with the host group, the host group did not adopt the mover's knowledge, even when it was superior to its own and would have improved its performance.

The experiment demonstrated the crucial importance of social identity for knowledge transfer: By extension, it can be argued that the transfer of tacit knowledge is closely related to social capital developed by a migrant within his/her host and domestic communities. The greater the sense of shared social identity, the easier it is to transfer knowledge.

Source: after Kane *et al.* (2005)

Box 3.7 **The diffusion of innovations**

Each move by a migrant – whether as emigration or return migration – has potential for learning, adoption and diffusion of new ideas among diverse societies. However, not all migration results in significant knowledge exchange. One explanation of how innovative ideas are adopted and diffused in diverse social systems is provided by Rogers' classical work on the diffusion of innovation (2003, first published in 1962).

Diffusion is the process by which an innovation is communicated over time, through certain channels, among the members of a social system. Each member of the system makes his or her decision on adopting an innovation by following a five-step process. In the (a) Knowledge Phase, a person or a group learns about an innovation and how it functions. In the (b) Persuasion Phase, favourable or unfavourable attitudes toward the innovation are formed via interaction with other group members. In the (c) Decision Phase, additional information on innovation is sought and a final decision is made whether to adopt or reject the innovation. In the (d) Implementation Phase, the innovation is put into regular use. Finally, in the (e) Confirmation Phase, the innovation is evaluated and either confirmed as justified or rejected.

The more radical the innovation, the more risky is its adoption. As most individuals are risk-averse, personal traits play a major role in decisions to adopt the innovation. Rogers found that the readiness of a member of a system to adopt a new idea, relative to other members, can be represented by a bell-shaped curve. He established five groups of members, among whom the process of ideas adoption may resemble the domino effect:

(1) The *innovators* represent 2.5 per cent of the system's members. These tend to be venturesome, educated individuals, with a greater propensity to take risk.
(2) The *early adopters* (13.5 per cent) evaluate the innovators' ideas and, if they deem these effective, may also adopt them. Early adopters tend to be popular and trusted opinion leaders whom the broader public follows when making important decisions about adopting new ideas, innovations or technologies. They set the path for:
(3) The *early majority* group (34.0 per cent), who are deliberate members of the system with large numbers of informal contacts. The domino effect continues and the adoption of new ideas becomes inevitable for:
(4) The sceptical and traditional-based members of the *late majority* (34.0 per cent) and:
(5) The cautious and fearful *laggards* (16.0 per cent).

Rogers' categories are purely statistical ones. The early and late majorities, for example, make up the core 68 per cent of the curve as defined by the first standard deviation. Each adoption curve refers to specific social dynamics surrounding a specific innovation. The same individuals may have different adoption curves in different circumstances and may drift from category to category for different innovations. Rogers' categories are also only meaningful if the innovations are adopted by all the relevant population. If only a few follow the innovator, there is no innovation diffusion.

Opinion leaders – who are strongest candidates for the role of early adopters – may favour a specific innovation, and encourage its rapid diffusion via the mass media. Frequent and strong personal contacts, on the other hand, may be more efficient when changing incumbent, or establishing new, firm attitudes. Different systems may account for different forms of innovation diffusion. Heterophilous social systems consist of members with diverse social and cultural backgrounds. They are more exposed to new ideas and prone to changes of routine. The promoters of change can concentrate on a relatively narrow elite of opinion leaders. Once this elite adopts an innovation, the domino effect supports diffusion of innovation via imitation. Homophilous social systems, on the other hand, consist of members with similar backgrounds. These systems tend to be more conservative and norm-abiding and their opinion leaders are less open to innovations. Diffusion of innovation is usually more difficult in the face of greater resistance. Promoters of change in such systems have to target larger numbers of opinion leaders and overcome their conservative stances. One useful strategy is to persuade the opinion leaders that the innovation in question is compatible with existing norms as this may confirm their leadership roles.

How do these ideas apply to our central focus on migrants? By extension, we can argue that the diffusion of innovations theory suggests that migrants arriving at host communities, and returnees coming back to their original societies, are more likely to enable knowledge adoption and transfer if these communities are heterophilous rather than homophilous. The intensity of inter-personal contacts, and good relations with opinion leaders, also seems to be essential, particularly for tacit knowledge transfers.

Source: after Rogers (2003)

Trust, as a determinant of knowledge transfer, is also important within groups. Group members have different status according to their appearance, education, gender, ethnicity, but also their social connections and expertise. Individuals with rich social networks tend to acquire more social capital and can activate more human and intellectual resources within the organization than members with less extensive networks. Research on in-group dynamics has demonstrated that socially-connected individuals assume greater similarity of opinion with those they are socially connected to than with those from whom they are socially isolated (Thomas-Hunt *et al.* 2003). Social isolates may be more likely to make unique knowledge contributions to the group, but they tend to be evaluated less favourably than socially-connected members. Social isolates, on the other hand, may defy the bias they face by gaining expert status and emphasizing their unique knowledge.

Securing adoption of a new idea, even when it has obvious advantages, is often problematic for migrants. Because of the complex interchange between the psychological factors affecting knowledge transfer and the institutional environment in the place where knowledge transfer is taking place, many ideas require lengthy periods (often several years) before they are adopted. Referring to the classic work of Rogers (2003) on the diffusion of innovations (see Box 3.6), we can identify further potential obstacles to the transfer of knowledge via migration. Trying to quickly convince new colleagues en masse of the value of their knowledge is generally fruitless, as the bell-shaped curve of innovation adoption suggests. The adoption of an innovation depends on awareness, interest, evaluation, trial and adoption by particular groups. It, therefore, makes more sense to start by convincing *innovators* and *early adopters*. Spreading ideas through society follows an S-curve, as the early adopters select the ideas initially, followed by the majority, until the idea becomes common. The speed of the adoption, of course, is influenced by many factors. These include, for example: (1) the speed at which adoption takes off and the speed at which later growth occurs; (2) the occurrence of competing ideas; and (3) path dependence generated by inertia in socio-cultural and economic institutions, which may lock certain ideas and technologies in place.

In addition, the type of knowledge to be transferred also mediates migrants' experiences of this, as can be seen in terms of Blackler's typology of knowledge, discussed earlier in this chapter. Embrained and embodied knowledge pose specific knowledge transfer challenges. They are related to an individual's mental and physical abilities. Knowledge obtained either via enlightenment (a 'brainwave') or via physical experience is difficult to transfer to others. Encultured and embedded knowledge are based on shared values and understanding, which may be place- or organization-specific, and can also be difficult to transfer to others. Chapter 2 has provided an initial discussion of the extent to which these different forms of knowledge can be transferred via migration.

Communities of practice and knowledge brokers

After a general discussion of learning and knowledge transfer, we turn to two channels or modes of knowledge transfer that are particularly significant for migrants: communities of practice and knowledge brokers. Our starting point is the proposition that tacit knowledge can be transferred both (i) within and between organizations sharing similar ideas, values and purposes and (ii) between organizations sharing different ideas, values and purposes. The former transfer is often realized via communities of practice, and the latter via knowledge brokers.

Communities of practice (CoP) are loose networks of individuals engaged in similar occupation, activities or interests. Wenger (1998: 215) identified human identity with learning and argued that learning is achieved via social participation:

> Because learning transforms who we are and what we can do, it is an experience of identity. It is not just an accumulation of skills and information, but also a process of becoming – to become a certain person or, conversely, to avoid becoming a certain person.

CoPs are informal associations that are established not via rules or design but via routines and repeated interaction. Wenger considered that CoPs have three main dimensions: (1) mutual engagement among the participants which sustains the community; (2) joint enterprise, which includes negotiation of diversity among members and the formation of a 'local' code of practice; and (3) a shared experience which reflects a history of mutual engagement.

The main purpose of a CoP is sharing ideas, finding solutions and accumulating stocks of collective knowledge. Most such communities associate individuals on a professional basis (for example, doctors, engineers or accountants). They exist parallel to the structures of firms and other organizations, but also across their boundaries. CoPs are usually single-disciplinary, or single-theme, communities but their strength is their capacity to link together those with diverse knowledge within particular parameters of shared ideas and values. CoPs are increasingly associated with knowledge management and are identified with developing social capital, nurturing new knowledge, stimulating innovation, and sharing existing tacit knowledge within or between organizations. The duration of CoPs transcends the lives or duration of specific public or private organizations, and the membership usually changes constantly as new members join or existing members exit. Many CoPs work on local basis (even within a particular organization) but others are geographically dispersed, because occupational similarities are at least as important as a local embeddedness, as exemplified by scientists.

Some authors argue that CoP are most effective in transferring tacit knowledge when they operate in a local context because many types of know-how and know-who are based on personal relationships that are strengthened by face-to-face contact. Zook (2004) exemplifies participative connection and multi-membership in a CoP of venture capitalists in the USA, who acquire and exchange tacit knowledge about new-technology firms, market conditions and financial institutions. These activities are not exclusively limited to local relationships, but are facilitated by physical proximity and are significantly assisted by sharing a communal social culture including collective beliefs, values, conventions and language.

Knowledge exchange between CoPs with non-local memberships is, of course, more difficult, due to potentially greater differences in socio-cultural milieus, or failures of shared understanding and values (encultured knowledge). Membership of distanciated CoPs, on the other hand, may offer greater potential for radical innovations, if it increases the potential to draw on different experiences, and knowledge. In his later writings, on what he termed expansiveness, Wenger (2000) examined this idea in terms of 'the breadth and scope' of identity. He concluded that 'a healthy identity' is not exclusively locally defined. Rather it will be constituted of multi-memberships, and will involve crossing multiple boundaries. Individuals with 'healthy' identities will actively seek out a broad range of experiences, and will be open to new learning possibilities.

Wenger's (1998) views on proximity are helpful. For him, CoPs are constellations of practices, understood as configurations of people that can be characterized by various notions of proximity, distance and locality. However, these notions are not necessarily congruent with physical proximity, institutional affiliations, or even frequency of interactions. While Wenger recognizes that spatial proximity can be important (for example, sharing offices or buildings), it does not necessarily create such communities. In this context, it can be argued that migration and other forms of geographical mobility can contribute to the evolution of CoPs in two important ways. First, migration may be critical for gaining entry to a localized CoP, with that membership being maintained after the migrant has returned home or relocated to some other place. Contacts and levels of trust may be maintained subsequently by return visits to the localized CoP, whether for manifestly professional or social purposes. This is perhaps exemplified by the movements of financial services workers between the global financial centres – a theme we return to in Chapter 6. Second, migration has potential to link two previously unrelated localized CoPs, via the movements of an individual who develops local knowledge and trust in both networks.

This last point takes us on to the notion of knowledge brokers and boundary spanners. Communities of practice and organizations have boundary exchanges with 'external others', and highly-skilled brokers play a key role in enabling such exchanges. Wenger (1998: 109) sees brokering as involving processes of translation, coordination and alignment between different perspectives. As discussed in Chapter 2, migration – or more specifically those migrants capable of reflexivity on the (especially embedded and encultured) knowledge that they carry with them – can be seen to play a particular role in knowledge brokering, understood in terms of the translation of different perspectives between different places. Wenger (2000: 223) explains this in terms of boundaries in general, although the comments apply equally well to international borders: they are 'areas of unusual learning, places where perspectives meet and new possibilities arise'. Radically new insights often arise at such boundaries, but it needs individuals with specific qualities to translate perspectives or ideas across these. These can be thought of as boundary spanners (Tushman and Scanlan 1981), individuals who can leverage external knowledge into organizations. They perform three key roles: accessing external knowledge, interpreting it, and refining it. As international borders still constitute significant barriers to acquiring external knowledge (see Chapter 5), migrants potentially have a distinctive role to play as boundary spanners. Brokers may operate in different ways. Following Tushman and Scanlan (1981), they can be boundary spanners who take care of one boundary at a time, 'roamers' who travel from place to place, and 'outposts' who bring back knowledge from 'the forefront': each of these can be related to particular forms of international migration and mobility (see Chapter 2).

Organizations and territorial economies can benefit from experts acting as knowledge brokers. Knowledge broking, however, depends on the individual broker's understanding of the knowledge in the recipient organization, and how it differs from the originating organization and/or country. This requires, as indicated earlier, a reflective understanding of encultered and embedded knowledge. Even seemingly 'acultural' sciences such as mathematics and astronomy are infused with cultural constructs that mediate the transfer of basic concepts between different cultures. The importance of shared socio-cultural understandings can be exemplified by three of the most illustrious migrants and knowledge brokers in history (Box 3.8).

***Box 3.8* Illustrious migrants and knowledge brokers**

Matteo Ricci (1552–1610) was a Jesuit monk, astronomer and mathematician. He was the first person to translate the Confucian classics into a Western language and, in turn, introduced Western science and technical concepts (for example, the automatic clock and a world atlas) to China. Ricci's great World Map brought about a revolution in traditional Chinese cosmography. In 1601 he presented himself at the Imperial court of Wanli. Ricci was able to rely on mathematical ideas that he had learnt from his teacher Clavius and, with the help of a Chinese mathematician, Xu Guangqi, translated the first parts of Euclid's *Elements* into Chinese. This is perhaps the first time that European mathematics and Chinese mathematics had interacted. Ricci also met a Korean emissary to China, Yi Su-gwang. Ricci's transmittance of western knowledge to Yi Su-gwang influenced and helped shape the foundation of the Silhak social reform movement in Korea (Cronin 2000).

Marco Polo (1254–1324) was a Venetian trader and explorer who, together with his father Niccolò and his uncle Maffeo, was one of the first Westerners to travel the Silk Road to China. Marco Polo spent some 25 years travelling in Asia on behalf of Mongol rulers. After returning from China in 1295, he wrote his book, *Il Milione*, which provided a lively description of the Mongol Empire and its riches. It became an instant success. The book was translated into many European languages and inspired Christopher Columbus's decision to try to reach the Far East by a western route. There are also claims that Marco Polo introduced to Europe some products from China, including paper money, ice cream and spaghetti (Larner 1999) – but irrespective of their veracity, he was a true knowledge broker.

Enrico Fermi (1901–54) was an Italian physicist who won the Nobel prize for his work on sub-atomic particles. In 1938, Mussolini's government forced Fermi to emigrate with his family to the USA. In August 1939 Albert Einstein sent his famous letter warning President Franklin D. Roosevelt of the probability that the Nazis were planning to build an atomic bomb. When Einstein and Szilárd overcame the administration's fears of foreigners being engaged in highly secret research, the American government started funding research on atomic energy. Fermi and his fellow scientists were the first to build a functional nuclear reactor in December 1942. Fermi's reactor was a landmark in the quest for nuclear energy, and would not have been possible without his brilliance, careful planning and meticulous calculations. Later Fermi and Szilárd's reactor work was incorporated into the Manhattan Project and helped prepare the way for the A-bomb (Fermi 1968).

The thirteenth-century traveller Marco Polo is a prime example of a flexible and highly-motivated business migrant, who also became one of the key knowledge brokers in history. Polo travelled Asia for a quarter of a century, entered Mongol administration and obtained an intimate knowledge of local customs, politics, economies and geography. He played an important role in transferring western ideas and science to China, but his accounts of China were met with deep scepticism after his return to Europe, because they represented such 'uncommon knowledge'. Even a display of the wealth that he had accumulated ('pockets full of precious gems') did not convince all the sceptics, because of the inherent difficulties in verifying his tales. Consequently, the knowledge that Marco Polo transferred from the Far East had only limited impact initially. It took about 80 years for the geographical knowledge Marco Polo had collected even to make an appearance in Europe's rudimentary maps.

Few people were better prepared for the role of knowledge broker than the Italian monk Matteo Ricci. Before sailing to the Ming Empire in China, he had received training in astronomy and mathematics in sixteenth-century Europe. After arriving in Macau, on the east coast of China, in 1582, Ricci learned Chinese and developed a deep understanding of Chinese culture, adapted to local habits and became one of the few Western scholars who had mastered Chinese classical script. He really did stand astride two different, and previously relatively unconnected knowledge systems, in a way that few others had managed to do so previously or since. The differences in culture and intellectual thought between Europe and China were enormous and it even took Ricci 19 years before the Chinese political and intellectual elites accepted his work. When he first attempted to visit Peking in 1595, he found the city closed to foreigners and instead settled in Nanking, before being allowed to enter Peking as late as 1601, which became his home until his death nine years later.

In contrast, Enrico Fermi and his fellow scientists enjoyed the advantages of broadly similar social, cultural and economic backgrounds when they migrated in the late 1930s to the USA. They were able to rely on communities of scientific practice, which strongly influenced their acquisition of visas and their place of resettlement. The US scientific community helped to persuade an American Government, suspicious of left-leaning intellectuals from Europe, that these European migrants could provide an invaluable contribution to nuclear research, The flexibility of the American universities, which welcomed the European migrants, enabled fruitful cooperation amongst scientists from different countries and from (somewhat) different education systems. No nuclear energy could have been released before the end of the Second World War, had the cream of European nuclear science not migrated to the USA. Their scientific advances, however, would probably not have been possible had the European scientists remained in their individual countries, instead of working together in the USA and drawing on the vast resources of its economy and educational system (Fermi 1968).

Conclusions: knowledge and the role of migrants

Knowledge is an elusive concept, and is difficult to disentangle from intuition, imagination, values and different forms of thought and evaluation. In early history, the transfer of knowledge was dependent on face-to-face contact, and so was geographically constrained until the invention of writing allowed some knowledge to be codified. After that it became much easier to disseminate knowledge across international borders. However, access to such knowledge still remained socially limited because of high rates of literacy, while many forms of tacit knowledge could not be codified. It was not until transport and communications revolutions, particularly in the nineteenth century, that many of the constraints on knowledge diffusion could be overcome. These had two apparently contradictory effects. First, transport changes facilitated human mobility and the transfer of tacit knowledge. Second, electronic communication, from the telephone age onwards, and the internet enabled some transfer of tacit knowledge at a distance.

Rather than seeing these as contradictory, they are perhaps better viewed as being complementary. A review of learning methods demonstrated that some forms of learning do demand physical proximity or relocation to be effective, notably learning by observation or learning in depth about encultured or embedded knowledge. For these reasons, migration remains necessary for some forms of knowledge transfer, and is an important aid to many other forms of knowledge transfer, including electronic, where building trust and networks are important. Given the continuing importance of international borders in delimiting differentiated knowledge systems, migrants potentially can play important roles as knowledge brokers or boundary spanners. Before proceeding to consider these roles in more detail, we turn in the following chapter to an assessment of the changing nature of international migration.

4 The changing context of international migration

A world on the move? International migration in perspective

Although debates about globalization and international migration sometimes give the impression of a world where everyone is on the move and crossing borders, this is far from reality. We do not know the exact numbers who move but, as Salt (2003: 5) comments: 'What is striking about these numbers is not how many people choose (or are able to choose) to live in another country, but how few'.

There are an estimated 200 million international migrants (GCIM 2005). While only a small fraction (3 per cent) of the total world population, they do constitute a significant part of the population increase in many world regions. Whereas, there were 48 countries where migrants accounted for at least 10 per cent of the population in 1970, by 2000 there were 70 such countries. In the period 1990–2000 international migrations generated 56 per cent of the population growth in the developed countries, and a remarkable 75 per cent in the USA and 89 per cent in Europe. Moreover, Europe's population would have declined by 4.4 million in the absence of international migration. However, it is still the case that only 3 per cent of the world's population live outside the country of their birth (GCIM 2005).

As can be seen from these impressive statistics, the overwhelming majority of people live in their countries of birth, and their movements (other than short-term sojourns) have been largely nationally bounded. However, there have been changes over time in the scale and relative importance of international migration, particularly in response to changes in the global organization of production. Goss and Lindquist (1995: 135) express this in terms of the extent to which individual lives intersect with changes in the organization and relationships of production.

Global economic growth, combined with changes in technology and the opening up of new territories for settlement, have led to significant international flows of migration in the nineteenth and twentieth centuries (see Box 1.2). Khadria (1999: 34), for example, explains that in the USA: 'Immigration reached a high in the 1890s when 10.4 out of every 1,000 US residents were immigrants. This rate dropped to below 1 in the 1930s and 1940s. It picked up again to reach a high of 4.7 in the 1990s.' New spatial divisions of labour emerged which both shaped and were shaped by these large-scale international migrations. In more recent decades these changes have been intensified by neo-liberalist tendencies in the world economy (Pellerin 1999: 476) whereby 'Migration movements constitute both a reaction to dislocations that transformations in the spatial and sectoral organization of the economy entail, and a constitutive element of the neo-liberal restructuring of capitalism'. It is important however not to fall into the trap of a simplistic association between migration and spatial divisions of labour, not least because, as Findlay and Li (1997) remind us, there may exist not one but several 'new international divisions of labour' that are interleaved and partially overlapping.

Without doubt there have been major changes in international migration in the late twentieth and early twenty-first centuries. The number of international migrants has approximately doubled in the last 25 years. This has been accompanied by marked change in the geography of international migration (Massey *et al.* 1993: 431).

> In traditional immigrant-receiving societies such as Australia, Canada, and the United States, the volume of immigration has grown and its composition has shifted decisively away from Europe, the historically dominant source, toward Asia, Africa, and Latin America. In Europe, meanwhile, countries that for centuries had been sending out migrants were suddenly transformed into immigrant-receiving societies. After 1945, virtually all countries in Western Europe began to attract significant numbers of workers from abroad. Although the migrants were initially drawn mainly from southern Europe, by the late 1960s they mostly came from developing countries in Africa, Asia, the Caribbean, and the Middle East.

These were not the only changes in international migration in this period, and there have also been significant flows of migration to some countries in the Middle East, as well as within Latin America, Africa and Asia. The current picture of international migration is summarized in Box 4.1, which reveals a step change in migration. While the relative proportion of the world's population who are now migrants may not be significantly different from the early twentieth century, the volume and its composition have changed markedly. There have been shifts in the duration and permanence of migration, with increases in return, serial and cyclical migration, all of which are considered later in this chapter. Moreover, there have also been changes in the skills, ages, and family status of migrants, as well as in motivations.

Box 4.1 **The global distribution of international migrants, 1960–2005**

The numbers of international migrants increased more than twofold in the period 1960–2005. The greatest increases were in Europe and Northern America, which became prime migration destinations (Figure 4.1): their shares of the total number of immigrants rose from 18.9 per cent and 16.6 per cent respectively in 1960, to 33.6 per cent and 23.3 per cent in 2005. These regions also accounted for the highest relative increases in the migrant population. In the same period the shares of total migrants in the total population increased from 3.4 per cent to 8.8 per cent in Europe and from 6.1 per cent to 13.5 per cent in North America (Figure 4.2). Most migration flows to Europe and North America were generated by income differences between the countries of origin and destination. In Asia and Africa, on the other hand, 'forced or semi-forced' migration was of considerable importance, due to famine, wars, drought and other push factors.

Source: after UN Population Division (2005)

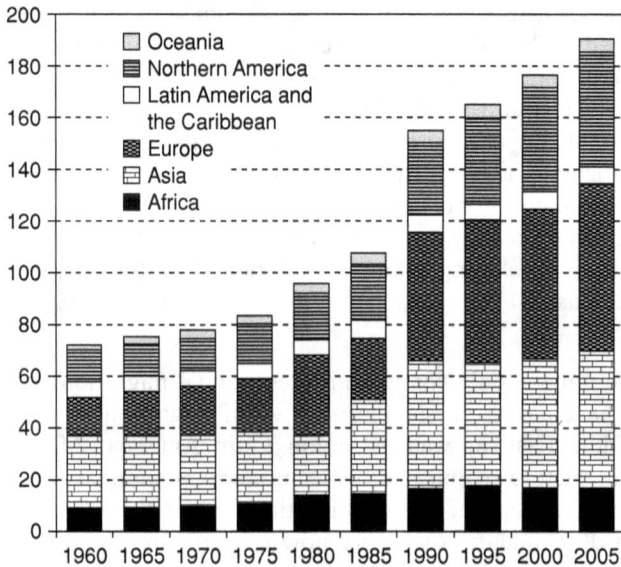

Figure 4.1 Estimated number of international migrants in macro world regions, 1960–2005 (millions in mid-year)

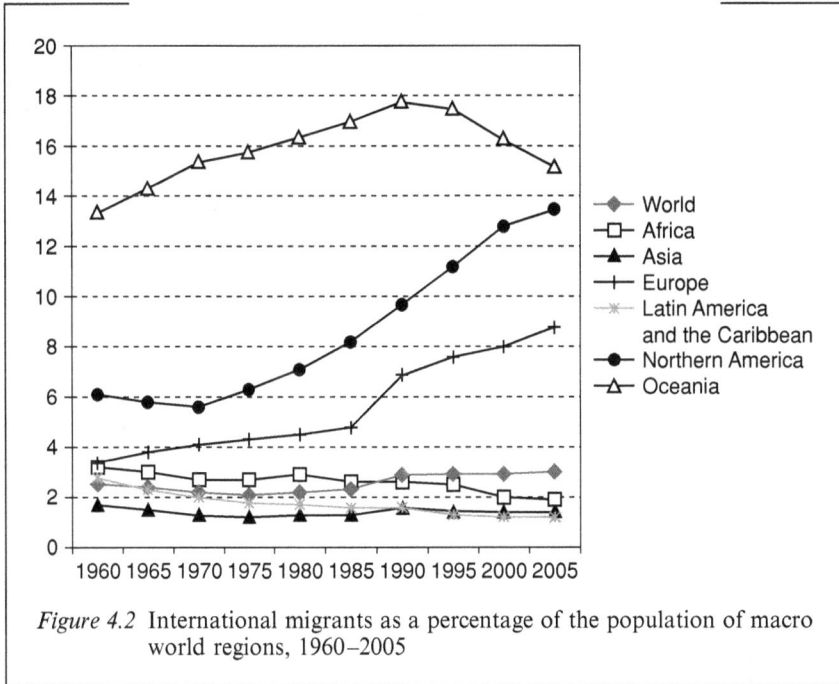

Figure 4.2 International migrants as a percentage of the population of macro world regions, 1960–2005

A review of global migration (United Nations 2004a: 77) emphasized that one of the most marked changes has been an increase in skilled international migration. In part this was a response to changes in the world economy and in part a response to the redrafting of national immigration systems in favour of 'skilled' as opposed to 'unskilled' migrants. Most of the world's leading economies have actively sought to attract skilled international migrants. This is reflected in the changing geography of migration, reinforcing flows between the more developed countries, and selectively between these and the less developed and emerging industrial economies. For example, it has led to a relative shift from migration to the USA in favour of Europe and Asia. There have also been changes in the complexity of migration, with increasing use being made of temporary visas, but also attempts made to attract foreign students and then encouraging them to stay on (see Chapter 5). Finally, there are different migration channels, and differences in the potential they provide for knowledge transfer, according to whether mobility is intra-, inter-, or extra-company (Williams 2006a) – discussed further in Chapter 6.

While governments, for both economic and political reasons, have emphasized policies that attract or at least tolerate skilled migration, this has not obviated the need for unskilled migrants in many economies. As the United Nations (2004a: 79) states: 'Although receiving countries have emphasized the need to attract highly skilled workers, population ageing and rising job expectations are also producing labour shortages in such low-skilled fields as agriculture, construction and domestic services.'

This is managed in different ways in different countries, including: the toleration, or at least the existence, of undocumented migration; the use of visas to fill short-term labour shortages in particular sectors such as farming, construction or tourism; allowing international students to work part-time; and schemes to permit working holidays for young people. Refugees and asylum seekers may also be given rights to work in some countries, and thereby add to the supplies of both skilled and unskilled migrant workers.

The growth in undocumented migration (United Nations 2004a: 82) has been a highly politicized issue, driven by increasingly restrictive immigration policies, and population displacements for a variety of reasons in the countries of origin. The channels used by undocumented migrants are varied, including reliance on traffickers, individual acts to cross borders without the required documents, or overstaying beyond the end of a visa period. Recent work (Ruhs and Anderson 2006) has emphasized that undocumented workers occupy a range, of often shifting, positions ranging from compliance to semi-compliance and non-compliance. Jasso and Rosenzweig (1982) also estimated that in the USA some 30 per cent of legal migrants had previously had some experience of being illegal. Many governments have tolerated undocumented immigration, either because of difficulties in enforcing regulations, or in recognition of labour shortages and the existence of 'difficult to fill' jobs.

Finally, although we are mainly concerned here with labour migration, the main source of immigration in many countries is family reunification. As the United Nations (2004a: 80) emphasizes, there is an apparent paradox between increasingly restrictive immigration policies and continuing growth in the foreign labour force, which in part stems from the sustained migration of family members. Some family members (particularly spouses) may enter the labour market immediately as soon as their entry requirements allow for this, while others – for example, children – represent potential future labour market entrants. Elderly migrants, moving to join their children, may also make an important economic contribution, perhaps by working part-time, but more significantly by providing child care for grandchildren.

Family reunification is also related to the simultaneous migration of accompanying family members. This is highly gendered and women tend mostly to have been seen in the role of 'trailing spouses' who move as economically passive partners of male economic migrants. This was always a generalization, but there has been a shift over time with more women acting as independent economic migrants, whether single or married (Boyle 2002). In these instances, migration can be a source of tension within the family, especially where there are dual careers (Ackers 2004). Therefore, possessing or acquiring knowledge, which can stimulate labour migration, has profound implications for other family members. However, as Raghuram (2002: 1) points out, there are still very few studies that explore how changes in the skills of a primary migrant shape family migration.

Whatever the type of international migration, and the channel of migration, this has long been recognized as a significant source of economic transfers and growth. International migration can influence wage levels in different labour markets. It is also a source of remittances for the sending countries. For example, according to the World Bank there were 16 countries in 2007 where migrant remittances accounted for 15 per cent or more of economic output, ranging from 36 per cent in Tajikistan to 16 per cent in Albania. Official remittances to developing countries alone in 2006 were $221.3 billion (Ratha and Xu 2008).

Looking beyond financial transfers, our argument in this volume focuses on the role of migration as a conduit for knowledge transfer. The migration channel and the status of the migrant have enormous potential impacts on knowledge transactions. For example, intra-company mobility involves planned transfers of encultured and embedded knowledge, as well as embrained and embodied knowledge. In contrast, inter- and extra-company mobility are more likely to involve unplanned knowledge spillovers, or knowledge seepage. There is of course also the fundamental bifurcation between migrants who enter a country under the regulations for skilled and unskilled migrants. Not only does this determine their initial engagement in knowledge transactions, but it may also have a lasting impact on their career development, learning and use of knowledge (see Chapter 7). Similarly, whether migrants are documented or undocumented, compliant or semicompliant, will influence their access to learning and skills, and their ability to apply successfully for particular types of jobs. For example, Matthews and Ruhs (2007) emphasize that employers in hospitality were likely to employ undocumented migrants in 'back-stage' jobs in the kitchen rather than front-stage jobs. In contrast, students and au pairs usually study or work under visa regulations that specifically recognize and encourage learning.

Family status is important. The partners and children of labour migrants also represent potential knowledgeable migrants, with knowledge of particular cultures and languages, or specialist embrained and embodied knowledge. Hence it is important to be aware of the constraints placed on the use of the knowledge by accompanying family members – individual countries have different policies with regard to whether, and how quickly, they can take jobs, and therefore use existing, or acquire new, knowledge (Hardill 2004). As Ackers (2004: 191) states: 'The tenacity of the "tied mover" pattern has a detrimental impact upon the career of female partners in dual-career households' but – depending on the regulatory framework – migration also creates new opportunities for them and for territorial economies.

Finally, it is important to emphasize that although we have focused on the national level, mainly because this is the key site in the regulation of international migration, migrants' work and knowledge transaction experiences are shaped in particular locales. There are, of course, concentrations of migrants in rural areas, particularly where they provide agricultural labour, but metropolitan areas are the focus of migration in most countries (see Box 4.2), and this is where the potential is greatest for utilizing the creativity and full range of knowledge transfer inherent in migration (see Chapter 5).

Box 4.2 The foreign-born population of selected major metropolitan areas

There are 25 major metropolitan areas where at least one-quarter of the population is foreign-born. Benton-Short *et al.* (2005) divide these into the 'usual suspects', that is, major and dynamic centres of economic activity such as Miami, Amsterdam, Toronto, Vancouver, Los Angeles, New York, Sydney, Frankfurt, London and Brussels. Although, to some extent, these are located in countries with long histories of in-migration, the proportions of the population who are foreign-born are far greater than in these countries in total, signifying a separate metropolitan effect. For example, while 11.5 per cent of the population of the USA is foreign-born, the proportion in Miami is 51 per cent.

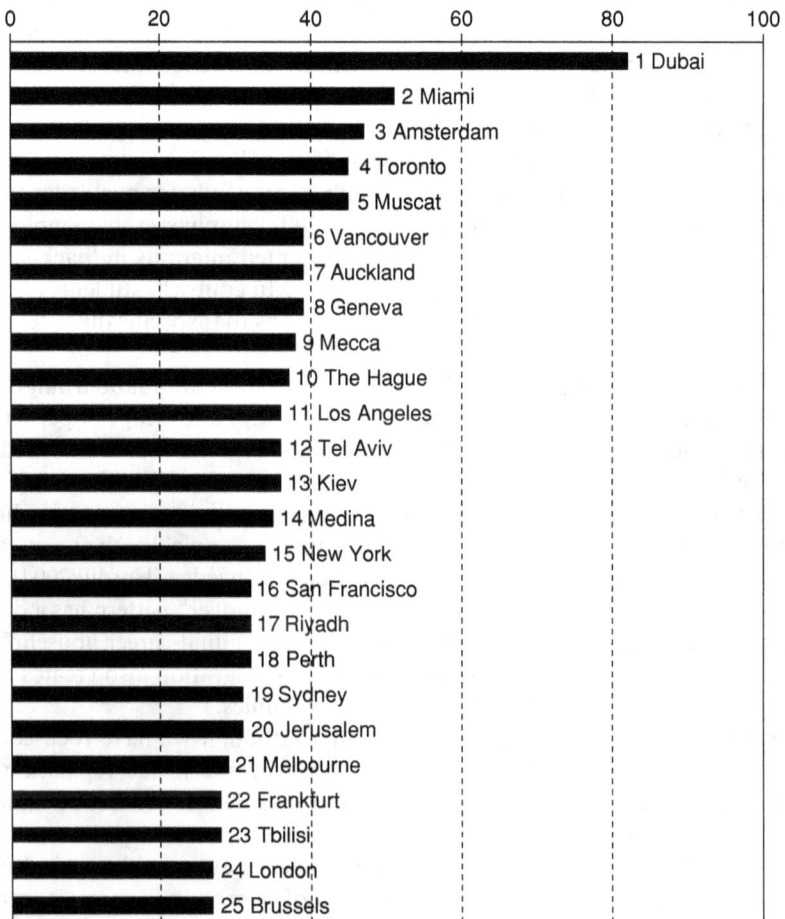

Figure 4.3 Foreign-born population: top 25 ranked metropolitan areas

There are also what Benton-Short *et al.* term 'unlikely suspects'. Seven of the top-ranked 25 cities are in the Middle East: Dubai, Muscat, Mecca, Tel Aviv, Medina, Riyadh and Jerusalem. Very few, and probably none, of these cities figure in the conventional rankings of the leading global cities, based on economic criteria, but they have all experienced rapid growth and, excepting Tel Aviv, are located in oil rich economies.

The most notable absentee from the table is Tokyo, usually considered one of the top-tier global cities. Despite its considerable economic strength, only 2.4 per cent of its population is foreign-born according to official data. Two other major Asian cities – Seoul and Jakarta – have even lower rankings, and the same applies to São Paulo and Mexico City in Latin America.

The absolute figures present a different picture but the percentage data do indicate that although international migration is an important feature of many of the world's leading metropolitan areas, this is not universally so.

Source: after Benton-Short *et al.* (2005)

The drivers of modern migration: demography, polity, economy and culture

The traditional distinctions between countries of origin, transit and destination are no longer valid in the face of modern migration trends, as most countries fall into all three categories although to different extents. International migrants now travel among countries at every level of economic development, and with almost all cultural and ideological backgrounds. However, as indicated earlier in this chapter, most migration flows follow the traditional 'south–north' route from developing to developed countries. According to GCIM (2005), the developed countries account for some 60 per cent of the total stock of international migrants. Nevertheless, 'south–south' flows between developing countries are also important and these countries account for some 40 per cent of the total migrant stock.

What are the major determinants of these massive migration flows? The largest population movements in human history have stemmed from the forces of nature and politics, rather than economics. Climate change, famine and drought, on the one hand, and political conflicts, ethnic cleansings and wars, on the other hand, can and have resulted in the 'movement of nations' in extremis, and in large numbers of people migrating throughout human history. These are usually categorized as forced migration. In addition, there has been migration fuelled by the pursuit of economic, ideological and cultural goals, which are usually considered to depend on voluntary decisions by migrants, although the borders between forced and voluntary migration are often blurred. Focusing here only on voluntary migration (while acknowledging this to be a problematic term), three main determinants can be identified: economics, demography and socio-cultural factors.

Economic determinants. While rates of poverty have been falling in most developing countries, income disparities between the developed and developing world have increased, especially if the rapidly growing emerging industrial economies are excluded. Per capita GDP was 41 times higher in the group of the high-income countries than in the low-income ones in 1971. By 2005 this had increased to a 66-fold difference. Migrants from poor to rich countries can earn nominal incomes that are 20–30 times higher than in their home countries, and this is a major influence on 'south–north' flows. Poverty itself, of course, is not the only determinant of migration, as is evidenced by the fact that only a small proportion of the populations of most low-income countries become migrants. Account also has to be taken of the costs and risks associated with migration, as well as the barriers posed by national regulation of immigration.

The flows between the more developed countries closely parallel the trade flows and capital flows between them, reflecting increasing global economic integration. There is not the same compelling poverty to drive migration, or the same massive gap in living standards. But there are differences in wages within the developed world and, perhaps more importantly, there are also considerable differences in the potential to develop careers in different national economic spaces. Moreover, international mobility has become integral to the operations of many transnational companies, being closely linked to knowledge transfers (see Chapter 6).

Demographic determinants. High-income countries have experienced a demographic transition, while the populations of most low-income countries continue to grow strongly. On average 5.4 children are born to each woman in Sub-Saharan Africa, compared with: 3.8 in the Arab World; 2.5 in Latin America and the Caribbean, and 1.4 in Europe. This has two implications. First, there is a massive demographic push factor in most less developed countries, as population growth exceeds the growth of jobs or the availability of land. Second, there is a growing demand for cheap and flexible migrant labour in the developed countries, to compensate for the diminishing size of their working-age populations, or at least the diminution of the ratio of economically-active to inactive persons. In concrete terms, migration provides the labour that maintains basic infrastructures and services, especially care services, as well as underpinning the pensions systems. It also – and this is often neglected – serves to strengthen the knowledge stocks of the more developed countries.

Cultural and lifestyle determinants. The increasing migration among the more developed countries has been informed by increasing living standards, rapid advances in transport and communication technologies, and changes in life expectancy, especially after retirement. The last contributes to significant later-life migration, which has three main forms: a return to countries of birth or childhood, family reunification as parents follow their children, and amenity-led migration, typified by the moves of Canadians to Florida, or British and Germans to Spain (Warnes and Williams 2006). There has also been the growth of youth, or young adult, migration, typified by student migration, 'rites of passage' migration such as the Big OE (overseas experience) migration from New Zealand and Australia to Europe, and various forms of tourism-work mobility (see Williams and Hall 2002).

The above account only provides the most generalized summary of the determinants of international migration. More nuanced accounts require more detailed case studies that we do not have space for in this volume. However, we can draw on migration theories for further insights, and here we largely follow the theoretical review provided by Massey *et al.* (1993). In brief, neo-classical economics focuses on macro-economic theories and wage differentials, that is broadly reflecting the first of our three main determinants. Similarly, human capital theories (see Chapter 2) focus on migration decisions as outcomes of the evaluations of the costs and benefits of migration, taking into account risk considerations. These also broadly fit into the first of the major determinants.

Two other sets of theories provide a more sophisticated account of these economic influences. First, the 'new economics of migration' emphasizes that migration decisions are usually taken collectively by families or other social units. This spreads the risks and the benefits across a larger group of people than the individual migrants. Second, the Dual Labour Market Theory argues that the employment – and, we could add, the knowledge and learning experiences – of migrants are shaped by labour market segmentation (see Chapter 2). It also argues that migration is driven more by pull factors in the destinations – the demand for certain types of low-cost, flexible labour – than by push factors, whether economic or demographic.

None of these theories provides an adequate explanation of the levels or the geographies of international migration. However, Network Theory (for example, Taylor 1986) is particularly useful in explaining how networks of inter-personal ties increase the probability of migration by lowering the costs and risks of mobility; they also explain the preference for particular destination countries (Bauer and Zimmermann 1997). Once a critical mass of migration is reached, costs are significantly reduced – in terms of finding jobs or accommodation, or overcoming lack of encultured knowledge – so that there is an element of cumulative causation in the migration flows. A number of other theories can also be drawn on to explain migration either in aggregate or at the individual level, particularly behavioural theories and those that emphasize values and identities (Brettel 2000). However, as already emphasized, these lie outside the scope of this book.

The determinants of migration are of critical importance in mediating knowledge and learning outcomes. These are far more likely to be positive where they are planned, coordinated and economic in nature, as opposed to being driven by political conflicts or natural calamities. They are also more likely to be economically significant among younger than later-life migrants, although the latter do contribute to consumption and some may continue to be economically productive. The influence of labour market segmentation on knowledge and learning has already been discussed in this context (Chapter 2), but this can be extended by reference to social networking. Migrants who move early in the migration chain are faced by far greater obstacles to learning than those who follow them. Later migrants may find that while strong networks may facilitate their movement, it may make it more difficult to acquire the encultured and embedded knowledge that allows them to progress their careers, especially if they become locked into jobs in the enclave labour market segment.

Changing forms of migration and mobility

While changes in the role of skills and knowledge in the economy are important drivers of migration, the changing nature of migration also shapes international economic transfers. There have been substantial shifts in the spatialities of migration, summarized by Meyer *et al.* (2001: 309) as becoming 'multilateral and polycentric though not completely multidirectional'. Several attempts have been made to try and capture some of the salient features of the changing forms of migration (see Koser and Salt 1997).

Logan (1992) proposed a typology which distinguished between moves intra and inter the less developed and more developed countries, but this did little to address the complexities of modern migration. Somewhat earlier, Gould (1988) had suggested a fourfold typology, which combined intra and inter North and South migration, with a distinction between permanent and circulatory movements. This was significant in emphasizing forms of migration other than permanent or long-duration moves.

Some approaches have emphasized organizational differences. One of the simplest of these, but highly significant for this book, is the distinction between inter- and intra-company mobility (Salt and Findlay 1989), because it provides a link to the way that knowledge systems are understood to exist within and beyond the boundaries of firms. Yet other typologies have focused on the distinctive mobilities associated with particular occupations. For example, Salt (1997) differentiated between:

- corporate transferees;
- technicians/visiting 'firemen';
- professionals;
- project specialists;
- consultant specialists;
- private career development and training movers;
- clergy and missionaries;
- entertainers, sportspeople and artists;
- business people and the independently wealthy;

- academics, including researchers and students;
- military personnel.

This is a problematic typology in conceptual terms because of the way it intermingles occupations, industries and roles, and inter- and intra-company mobility. Nevertheless, it does succeed in identifying groups of workers with distinctive migration and mobility patterns. An alternative occupational categorization is provided by Mahroum (1999), who identifies five main types of skilled labour migrants:

- professional and managerial;
- engineers and technicians;
- academics and scientists;
- entrepreneurs; and
- students.

While useful in highlighting the focus of much existing research on a relatively narrow range of occupations (primarily the professional and managerial) this is also somewhat arbitrary in interleaving occupations and employment status. Here we limit the discussion to just three types of migration, defined in terms of spatiality and duration: the shift from 'permanent' to temporary migration; the significance of return migration; and the increasing prominence of serial migration and circulation.

From permanent to temporary migration

There have been major changes in how international migration is conceptualized. In particular, there has been a shift from permanent to temporary migration, with the implication that we need to think about the full cycle of migration, or possibly cycles of migration. This has posed a challenge for theories that were constructed around the notion of migration being a one-off event, leading to settlement, the gradual but slow acquisition of some degree of local knowledge, and social integration either with the migrant or the indigenous community. Initially this was particularly challenging for human capital theories but Dustmann and Weiss (2007) explain some of the ways in which return migration can be explained within this theoretical framework (see also Chapter 2). Most importantly for our purposes, they argue that the human capital that migrants have acquired from working abroad can result not only in higher wages in the destination, but also in the origin country, especially if they have acquired particular technical, managerial or language skills.

There is still a systematic lack of data on the duration of migration. However, drawing on a relatively small purposive sample, Baláž *et al.* (2004) found that the average stay for au pairs, students and even professional and managerial returnees from the UK to Slovakia were in the range of 3–9 months. Putting aside the debates as to how to differentiate migration from other types of mobility, and also the fact they did not study those who had decided not to return to Slovakia, the findings do indicate the importance of shorter-term migration for at least some types of migrants. This study also found that young, well-educated migrants have steep learning curves in terms of accumulating knowledge, and they return home with significantly enhanced competences, many of which reflect new embrained and encultured knowledge in particular. Hannerz (1996: 131) also captures the multi-faced meanings of such short- to medium-term sojourns, particular for what he terms young 'expressive specialists', who come to the world cities in particular 'because of unique learning opportunities' and the economic and cultural imperatives of simply 'being in the right place'.

Return migration

The corollary of the shift from permanent to temporary migration is the increasing importance of return migration. In a way they are indivisible, but we consider return migration separately here because return migration studies focus on economic behaviour post-return. This has long been a relatively neglected field, due to data deficiencies but also to the way in which most theories were constructed around the notion of 'permanent' migration. King (1986b) considers that there was virtually no significant research interest in the subject before the 1960s. More than a decade later, Ghosh (2000) was still able to refer to return migration as being an 'unwritten chapter' in migration studies.

This research lacuna is surprising because return has long been a feature for some participants in international migration, even of the long distance transatlantic migrations of the nineteenth century. However, it has increased sharply in recent decades as a result of the shift to temporary migration, and has been facilitated by changes in transport and communications. Hence, Newbold's (2001: 28) estimate that some 20–30 per cent of all recorded moves are returns is probably a significant under-estimate. This is indicated by Dustmann and Weiss's (2007) estimate for the UK, based on the Labour Force Survey: of those migrants who had been in the UK at least one year at the time of the survey, only 60 per cent of men and 45 per cent of women were still there after five years. This is also likely to be an under-estimate because of the one-year cut-off point they used in their analysis. Taking into account the shift to temporary migration or mobility, there is increasing evidence of relatively short-term international mobility involving students, intra-firm placements, au pairs, and many others – as indicated in Baláž *et al.*'s (2004) study of temporary migration from Slovakia to the UK in early 2002.

One of the key issues in understanding how migrants use the knowledge and capital they have acquired abroad is the reason for return. Cerase (1974) provided an early typology for the main reasons for return that still has resonance today. It should be noted that he mainly emphasized different experiences of acculturation, which in terms of our focus on knowledge relates most closely to the acquisition of encultured knowledge:

(1) Return of failure. Failure to find a job in host country, and or failure to cope, and so the process of integration never really began.
(2) Return of conservatism. This is a planned move that is typically effected when the migrant has saved enough to achieve his or her desired economic goals. The migrant has tended to remain orientated to the home country, and to have experienced relatively limited acculturation throughout the period spent abroad.
(3) Return of innovation. The migrant returns with the intention of being innovative, perhaps on realizing that he or she has reached the limits of what can be achieved in the destination with the newly acquired skills and knowledge.
(4) Return of retirement. This return, whether relatively early or in terms of later life, can encounter particular economic obstacles if the migrant has been away a long time, and is no longer in tune with local values and practices in the country of origin.

Each of these types of migration has different implications for economic behaviour and knowledge transactions. The return of failure is unlikely to have resulted in the acquisition of any useful forms of knowledge. The return of conservatism is likely to have resulted in a narrow range of new encultured and embedded knowledge. The return of retirement may be accompanied by varying degrees of potential knowledge transfer, but this is rarely economically significant because of their age and motivations. Hence the most important category is the return of innovation.

King (1986b: 18–20) further explored the empirical evidence for the return of innovation, and identified some of the conditions that mediated the effectiveness of innovative behaviour. They included the following:

• The number and spatial/temporal concentration of returnees because it requires a critical mass of new ideas and open-minded people to effect changes in a particular area.
• The duration of the migration. If it is too short, then the migrant will lack new ideas and capital, whereas if it is too long he or she may find that they have become alienated from the home society (in other words, they lack encultured knowledge).
• The social class of the migrant: the more educated that he or she is, then the greater will be their impact.
• The differences between countries and regions, especially the different opportunity structures provided by metropolitan and non-metropolitan areas.

- The nature of the training and skills (or, in our terms, the knowledge) they have acquired abroad.
- Whether their return is planned and organized or unplanned and spontaneous; the economic impact of the former can be expected to be greater.

Cerase (1974) was writing about the experiences of Italian migrants in the USA in the middle of the twentieth century, and his typology of return is therefore relatively narrow. King was also editing a book that mainly looked at the experiences of the age of mass migration. Not surprisingly, therefore, there are some problems in applying their conclusions to understanding the changing migration scene in the late twentieth century.

Although return migration, even the return of innovation, does not always have strong positive economic impacts on particular areas, commentators have been more positive about the return of skilled migrants. Khadria (1999) was one of the first to recognize that the return of skilled migrants could represent what he called 'silverstreaks', or positive stories about the economic consequences of migration. The central argument is that successful returnees come back with increased stocks of knowledge and financial capital. While most of the literature has concentrated on the role of remittances and savings in establishing small businesses, the enhanced advantage that returnees possess is likely to be as much due to the accumulation of knowledge and skills (see Box 4.3). And while these may be especially prized in the cases of scientists or top-professionals, similar arguments also apply to other groups of skilled migrants.

Box 4.3 **The return to return migration**

McCormick and Wahba (2001) explored a sample of 1,526 migrants, who returned to Egypt after a minimum stay of six months abroad. Among these, 18 per cent were university graduates, 23 per cent had remained at school beyond 14 years of age, and 30 per cent were illiterate.

International migration had a profound impact on their employment. While some 44 per cent had worked in the public sector before migration, only 9 per cent took jobs in this sector after returning. Instead, there was a significant increase in the proportion who were employers – from 10 to 19 per cent. This shift in employment status was particularly pronounced for young and educated employees in the public sector, who wanted to secure jobs in the private sector.

Of particular note is that McCormick and Wahba explored whether overseas employment enabled return migrants to develop businesses. If so, was this because it had been a source of accumulated savings or a means to acquire new skills and ideas? They found that the primary economic influence of the foreign experience among literates derived from the period of time spent working abroad (and the acquisition of skills), while the effects of accumulated savings were less important. In the case of illiterates, on the other hand, savings were more important for establishing businesses.

McCormick and Wahba's research provides broad support for the argument that the constraints on entrepreneurship in less developed countries derive not only from a general lack of capital and poor financial service infrastructures, but also from lower levels of human capital accumulation. Migrants who have already acquired a certain amount of human capital in their home countries are more likely to accumulate human capital while working abroad and to benefit from its transfer after they return.

This can be related to Ghosh's (2000) work which identified three conditions for countries to benefit from return migration:

(1) workers return with knowledge that is more advanced, or with better skills than they would have acquired at home;
(2) the knowledge and skills acquired abroad are relevant to the needs of the home country economy; and
(3) the migrants must be willing and have the opportunity to use the skills upon return.

As for highly-skilled migrants, Olesen (2002) suggests that the country of origin derives the maximum benefit when the migrants leave for relatively short periods of 10–15 years, and return with financial, human and social capital.

Source: after McCormick and Wahba (2001); Cerase (1974); Ghosh (2000); Olesen (2002)

There is also emerging evidence that short-term migration is undertaken by migrants in order to acquire knowledge, which they intend to commodify after returning – such as the au pairs and language students surveyed by Baláž *et al.* (2004) (see Box 4.4). Relatively short periods of residence abroad may allow the acquisition of particular competences (in networking, communication etc.), self-confidence, and reflective encultured and embedded knowledge. It also furnishes migrants with language skills, which in combination with other forms of knowledge, can attract a significant wage premium in the country of origin (Williams and Baláž 2005a). This accords with more recent thinking, in human capital theories, about the earnings function between wages and length of residence; while earnings are likely to increase over time in the destination, some migrants may be able to attract a considerable real income premium by returning to use their acquired human capital in the country of origin (Dustmann and Weiss 2007). Some of these complexities are recognized by Newbold (2001: 37–8), who makes links between the ages, motivations and human capital of the returnees.

Box 4.4 **Different stories of return: Slovakian experiences**

Williams and Baláž in a series of papers explored the experiences of returned migrants in Slovakia who had spent at least three months living in the UK in the early 2000s. Based on 186 in-depth interviews, they sought to assess the extent to which these periods of relatively short migration (6–12 months mostly) had been a means for three distinctive groups of migrants to add to their human capital. The three groups were professional and managerial workers, students and au pairs. Each group had entered the UK under a different visa arrangement which provided contrasting opportunities for formal employment and learning.

The main finding was that the migrants were overwhelmingly positive about their experiences in the UK. These were translated into real gains in either income or social status for more than one-third of each group (Table 4.1). This was particularly impressive in overall terms given that their sojourns to the UK had been relatively short, being indicative of steep learning curves. It was particularly noteworthy that many of the au pairs – whose work experience was formally tied to the home of their host families – recorded substantial income gains. The fact that fewer recorded improvements in their status reflected the fact that many were students who returned to complete their studies, and had not yet entered full-time employment in Slovakia.

Table 4.1 The influence of migration on the status and income of returnees in Slovakia

	Professionals	*Students*	*Au pairs*
% whose social status had improved	43.8	30.9	14.9
% whose income had improved	40.6	38.2	37.3

Source: Williams and Baláž (2005a)

Very few of the migrants had acquired formal qualifications during this period, except for the au pairs who often had obtained English-language proficiency certificates. Instead, what they valued most was learning English, which although of some economic value in the UK, was of considerable value in the job market in Slovakia, an emerging market economy where such skills attracted relatively high wages, and were a requirement for working with many companies. This was summed up by one migrant, Peter:

> I became an entrepreneur after my return from the UK. I needed English to communicate with clients, some of whom are foreign. I speak good English now, which is essential for that. I think, my social status also improved, because I learnt English.

Similarly, another returnee, Miriam, commented:

> I do use the English which I learnt . . . I am working as an inter-
> preter in a Czech firm. This is interesting and well-paid. I think I
> got the job because I can communicate fluently in English, solve
> problems, and rely on myself to deal with problems.

Miriam also highlighted the value placed on acquiring a range of
social competences, including communication skills, social networking and team-working. Migrants also returned with a greater sense
of self-confidence, based on having completed a successful period of
migration abroad. Róbert summed up some of the competences he
had acquired: 'The most valuable experience I obtained in the UK was
not so much professional skills – I could learn these anywhere – as the
ability to manage crises and to communicate.'

A similar account was provided by Marcela who had returned from
the UK to open a beauty salon, and who had no doubt as to the
beneficial effects of her time spent working as an au pair in the UK:

> I did not learn any particular skills in the UK that were helpful
> to my current business in Slovakia. Working in the UK, however,
> was essential for my life. I learned to be independent and am self-
> confident. If I had money, but no self-confidence, I would not have
> been able to start my own business.

Source: Baláž and Williams (2004); Williams and Baláž (2004, 2005a)

Encultured and embedded knowledge are also highly valued by the policy
makers who seek to attract highly-skilled migrants to return to less developed or emerging industrial economies. For example, Jonkers (2004: 1)
demonstrates that returned scientific migrants to China are valued because
they have 'institutional knowledge and skills specific to that system' – or what
can be termed embedded knowledge of the country of destination
and return (Box 4.5). In this context, we can also note that Meyer (2001)
argues that it is advantageous if there are 'critical mass densities' of
expatriates as this facilitates more frequent interactions and collective
endeavours.

Migration may become not only a means of acquiring knowledge, but
a recognizable symbol of expertise (much like a qualification) for the returnee. This commoditization of mobility experiences stands in stark
contrast to the traditional emphasis in human capital theories on the premium paid to nationally-specific education and knowledge, whether because
of the assumed learning curve that underpins the earnings function in human
capital theories, or because of interest group politics that defend the

---

Box 4.5 **The role of return migrants in the development of Beijing's biotechnology cluster**

The System of Innovation (SI) concept was used by Jonkers (2004) to identify the main organizations in the Chinese research system. One of the features of the SI literature is that it offers a broad perspective on the innovation system, emphasizing that learning often takes place outside the formal R&D context. Instead it stresses the importance of learning by doing, learning by interacting, and user–producer inter- actions (Lundvall and Johnson 1994).

Innovation in the biotechnology sector depends heavily on the scientific research that takes place in the (public and private) organizations that comprise the 'formal research system'. Because knowledge and skills in biotechnology have an important tacit (personal knowledge) com- ponent, the stocks of knowledge held by individuals are especially relevant for this sector. So too are the flows of restricted codified and tacit knowledge which are transferred within networks. Not surprisingly, therefore, in comparison to other natural sciences, a high degree of interpersonal research collaboration is characteristic of this sector.

Against this background, China has actively sought to encourage Chinese biotechnologists working abroad to return to China to pro- vide the human capital base for developing high-tech industries. China is estimated to have the largest number of researchers working abroad of any country in the world. Apart from transferring scientific know- ledge, the returnees are also valued because they provide a means to connect Chinese research to global R&D networks, thereby creating conduits for the inward flow of international knowledge which was considered important for scientific development in general, and for biotechnological innovation in particular.

Source: after Jonkers (2004)

---

interests of indigenous workers (Duvander 2001) and, we can add, 'domestic knowledge'.

Serial and circular migration

It is only a small step from recognizing the importance of return migration, to acknowledging that migrants may be involved in more than one act of migration. Indeed, the possibility of repeat migration, in various forms, was recognized by King (1986a) in his landmark review of return migration. In this section we consider two particular forms of this, serial and circular migration.

Ossman (2004) coined the term 'serial migration' to indicate migrants who undertook a series of migrations, rather than a simple act of migration, or migration and return. He notes that such migrants are absent from all the major statistical sources on migrants, not least because these tend to take snapshots that, at best, record a single origin and destination.

> Their citizenship or official residence may not correspond to the place where they actually live and work. They are neither travelers always on the go, nomads without homes, nor cosmopolitans whose blithe travels might take them anywhere. . . . Unlike diplomats or managers of multinationals who move across the world to work, they do not have a stable institutional or national reference point that makes sense of their displacements. Serial migrants can rarely find direction and stability by reference to a stable social or professional medium in which they can progressively design their lives. Their success in each new place depends on their ability to deal with ruptures at every level. They must repeatedly learn to find a place for themselves not only as cultural chameleons, but also as social actors.

This powerful description uses the metaphor of 'chameleons' for what is, in effect, the process of acquiring encultured knowledge in a sequence of different environments where socially-situated learning always poses new challenges. The most successful of the serial migrants have learnt to learn quickly and efficiently, that is having the 'ability to learn at every level'. This requires the ability to learn about new cultures quickly, but also to be effective social actors, that is drawing on a range of social competences, including networking abilities, in order to establish themselves in new settings. Such individuals may be invaluable to, for example, international companies that need a cadre of managers able to move on behalf of the company, setting up new branches or firefighting where problems emerge in existing branches. They are also likely to be invaluable to international consultancy firms. But they can also be free-agent labour migrants (Williams 2006a) moving outside of corporate frameworks.

Another form of increasingly important mobility is circular migration, where an individual circulates repeatedly between two places. This can take different forms, and can be what King (1986a: 10) terms 'periodic' migration, that is relatively short-term, at weekly or monthly intervals, or seasonal. Weekly and monthly circulation tends to blur into different forms of long-distance commuting. Seasonal migration is particularly characteristic of those sectors where production is distinctly seasonal, such as agriculture or tourism. An individual's persistent regular presence in two places makes possible the development of extensive social networks and experiences of social integration that are unavailable to the more 'butterfly-like' mobility of the serial migrant. Global seafarers are one of the more unusual groups of circular migrants (Box 4.6).

Box 4.6 **The global labour market for seafarers**

By the end of the twentieth century, there were clear signs of the globalization of trade and capital. The emergence of the global labour
market, on the other hand, was much slower and was hampered by a
large number of institutional obstacles. However, some occupations
did demonstrate signs of becoming increasingly globalized. Wu and
Morris (2006) analyzed one of these in their study of the emergence of
a global market for seafarers.

In the late 1970s, many ship owners in the more developed countries
switched their ships' registration from a national flag to a 'flag of convenience' (for example, to that of Panama or Liberia) in order to escape
the constraints of national regulations and to recruit low-cost labour. This
contributed to the emergence of a global labour market for seafarers.
As a consequence, the shares of seafarers from the advanced countries
in the total seafarer labour market have decreased, while the shares of
seafarers from transition and developing countries have increased.

The Seafarers' International Research Centre (SIRC) database, con-
taining information on nearly 100,000 seafarers working on board over
5,000 ocean-going vessels, allows for the comparison of the nationality of seafarers with their ship's 'country of economic benefit' (CEB).
The CEB identifies the ship's *de facto* national identity, because the
official 'flag of convenience' rarely reflects the nationality of its real owner.
Seafarers working for employers from different CEBs can, in some
respects, be considered 'global' seafarers while those employed by their
national fleets can be considered 'national'. Some 68.4 per cent of all
seafarers were 'global' by 2002 (see Figure 4.4). Moreover, 60 per cent
of all crews were composed of two or more nationalities, while single-
nationality crews accounted only for 40 per cent.

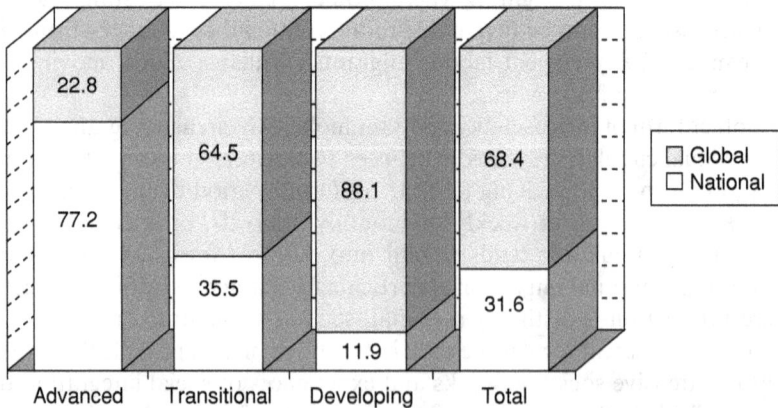

Figure 4.4 Division of global and national seafarers by region of origin (%)

The global labour market was a major employment channel for seafarers from transitional and developing economies. Demands on global seafarers and multinational crews have been fuelled by a shortage of low-wage seafarers. This created a window of opportunity for seafarers from Eastern Europe and China. Since the 1980s difficulties related to economic and social transitions in these countries have generated a large reserve of surplus labour. Unlike most occupations in Eastern Europe and China, seafarers had an opportunity to tap into a global labour market. As a result, they provide 47 per cent of the global seafarer labour. Wu and Morris also found evidence for some segmentation in the global seafarer market. China and Russia, for example, had large national fleets and their seafarers were more likely to be employed on their national ships. Poles and Ukrainians, on the other hand, were more likely to be employed in the global labour market for seafarers.

Source: after Wu and Morris (2006)

In different ways, both serial and circular migrants engage in what Duany (2002: 358) termed mobile livelihoods:

> By mobile livelihoods I mean the spatial extension of people's means of subsistence across various local, regional, and national settings. Circulation is only one, albeit an important, form of physical and social displacement that reflects mobile livelihood practices. People who frequently cross geopolitical frontiers also move along the edges of cultural borders, such as those created by language, citizenship, race, ethnicity and gender ideology. Thus the development of mobile livelihoods has serious implications for the construction of labor markets, discourses of citizenship, language policies, and national identities.

It also has implications for the capacity of these individuals to develop multiple embedded and encultured knowledges, which may be particularly valued by some employers, although this may be at a considerable cost to the individuals in terms of the amounts of time they have to devote to learning in multiple sites. We consider some of these issues further in the following section.

Diasporas and transnationalism

Diasporas and knowledge transfer

The term diaspora has its origins in ancient Greek, where it meant 'the scattering of seeds', but it came to refer particularly to the dispersal of Jews through much of modern history, until mid-twentieth-century social scientists adopted this as a generic term referring to the dispersal of any people. Not all migration flows become diasporas but these exist where there are persistent attachments, identifications and contacts with a global community who share a collective identity. Cohen's (1997: 175–6) influential book, *Global Diasporas: An Introduction*, considered that social networks were the 'flesh and blood' of the world economy:

> Many social and economic actors, including states, international organizations and transnational corporations, propel a world economy. These may be the sinews binding the ends of the earth together, but the flesh and blood are the family, kin, clan and ethnic networks that organize trade and allow the unencumbered flow of economic transactions and family migrants.

Where these networks involve communities settled outside their real or imagined natal territories, they constitute diasporas. Although these take many different forms, they all acknowledge an emotional tie to 'the old country'. There is, therefore, the acceptance of 'an inescapable link with their past migration history and a sense of co-ethnicity with others of a similar background' (Cohen 1997: ix). For our purposes, this means that diasporas constitute a distinctive medium for knowledge transfer. Cohen recognized different types of diasporas, including those centred on trade and labour. But all diasporas constitute a format for knowledge transfer that has potential implications for economic practices.

Diasporas bridge international boundaries with ties that are based on loyalty, emotion and a sense of shared history. They are founded on trust and shared identities that exclude others while binding the members together. For economic purposes, they can become conduits for the mobility of capital, trade and knowledge between countries, whether this means the 'old country', or the many different countries that they now live in around the world. Diasporas may function on an informal or individual basis, or may have nodal points such as hometown associations that provide nuclei for these dispersed international communities. The associations may be concerned with the welfare and practices of the diaspora in a particular region or country, or may actively involve themselves in transferring resources to the old country. This often takes the form of raising funds for particular projects such as a school, or a religious or community centre.

It is hardly surprising that policy makers should have seized upon diasporas as a potential instrument of economic policy. The United Nations (2004a: 122) *World Economic and Social Survey 2004* summarizes this perspective:

> Whether through direct contact or through the so-called hometown associations, the migrant diaspora can be a significant source of, and facilitator in, technology, knowledge, skill and financial transfers, as well as a facilitator of FDI in the home countries. The leveraging of diaspora

contacts and resources by policy makers, as well as by the private sector in developing countries, should be encouraged . . . with a view to increasing their development impact.

As indicated in this report, diasporas can play a role in the transfer of technology, knowledge, skills and remittances, as well as in facilitating foreign direct investment. In practice these are inter-related and all have implications in some way for knowledge transfer. Drawing on Mahroum *et al.* (2006: 28–9), we can identify three main channels through which diasporas can transfer knowledge:

- The most obvious channel is corporeal mobility, or physical return by skilled emigrants. The Hsinchu Science-Based Industrial Park in Taiwan provides one example of such a channel of knowledge transfer: returned migrants from Silicon Valley are responsible for half of the new-start companies at Hsinchu according to Lucas (2001).
- Members of the diaspora can transfer knowledge without relocating by means of distanciated communication, especially using ICT. Members of the diaspora may form professional networks that either intentionally or indirectly serve to transfer knowledge to the 'old country'. These can approximate to an ethnically- or nationally-selective version of Wenger's (1998) communities of practice. Given the limitations on the transfer of tacit knowledge by distanciated contacts, such networks are likely to be more effective if they also involve relatively frequent face-to-face contacts though short-term visits. Biao (2005: 51) provides an example of knowledge transfer in one such diaspora. Dr Zhang, a Chinese national working at a London university, had close contacts with a department in a leading university in Beijing, both with colleagues and friends. When the department imported two sophisticated items of equipment, there was uncertainty as to how to maximize their use. Dr Zhang demonstrated in person the capabilities of the equipment to them. This is 'knowledge which is taken for granted in Zhang's workplace in the UK, but would have been very difficult to learn in China without such personal connections'.
- Members of the diaspora are relatively more likely to invest in the 'old country', because they have privileged knowledge about it compared to most other countries (Lucas 2001). This is particularly marked in the case of India, as reported by Mahroum *et al.* (2006: 29):

  India's technology-oriented diaspora stand behind much of the FDI in the country's emerging technology hubs of Bangalore and Hyderabad. . . . Indeed, the Indian diaspora are the biggest investors in advanced medical services in India, a vibrant sector that caters largely to the Indian middle class, but also to many patients from overseas who bring foreign exchange and jobs to India. Indeed, some Indian health-care workers who had emigrated to

Europe and the United States have returned to India because of the opportunities and rewards of working in this private sector.

Interest in the knowledge transfer role of diasporas took off in the 1990s and was particularly strong in the international scientific community. Gamlen (2005: 21) considers that the 'emergence of homeland strategies to leverage diaspora knowledge networks' was first noted by Meyer *et al.* (1997) in a groundbreaking case study of the Colombian 'diaspora'. Interest in the phenomenon has since grown exponentially. Although referred to by various terms, such as scientific diasporas or knowledge networks abroad, Meyer and Wattiaux (2006: 4) recommend the use of the term 'diaspora knowledge networks' (DKN), which was originated by the International Committee for Social Science Information and Documentation for UNESCO. The DKN is characterized by a commitment among its members to use their skills to contribute to development in their natal country; increasingly they rely on a connectionist approach, where social capital is more important than corporeal return migration.

The activities of DKNs take many forms. Mahroum *et al.* (2006: 28) report the following in their review:

- exchange of scientific, technical, administrative or political information (e.g. the creation of the new Colombian National S&T system in the early 1990s, by prominent expatriates);
- specialist knowledge transfer (e.g. waste management procedures from the École Polytechnique Fédérale de Lausanne-Switzerland, with the Universidad del Valle, Cali-Colombia);
- 'scientific or technological diplomacy' or promoting the country in the R&D and business community of the countries of destination (e.g. South African medical research in the UK);
- joint projects, partly on a virtual basis;
- training: attending home-country sessions and hosting students abroad;
- enterprise creation and the possible return of expatriates on a part-time or permanent basis (Chinese high-tech firms with returnees in science parks);
- *ad-hoc* consultations, for example on research/development (peer review, job recruitment, technology assessment).

There is of course some exaggeration about the role of DKNs. On the one hand, they can influence where ideas are generated and also, more commonly, the diffusion of knowledge (Lucas 2001). On the other hand, however, Kapur (2001) questions the extent to which diasporas have a privileged knowledge of the old countries. Do they fully understand how knowledge or capital transfers will be mediated by local economic or cultural circumstances, or has their imagined view of the 'home' country become divorced from reality over time, perhaps over several generations? DKNs will be most effective in 'old countries' such as India or China, where there has already been considerable investment in the scientific infrastructure, and where there is the dynamic growth of an emergent industrial economy. The links between

Indian workers in Silicon Valley and India are often held to be another example of an effective DKN (Saxenian 2002), although Lucas (2001) cautions
that the availability of a substantial supply of local, low-cost skilled labour
may have been an even more significant factor in the growth of India's ICT
industries (see Box 4.7).

Box 4.7 **The discreet charm of the diaspora**

Diasporas play an important role in providing employment, establishing businesses, and the success of these businesses both in host and home
countries. Many diaspora members organize themselves in ethnic associations, which try to foster knowledge and business ties between the
countries of 'home' and origin. Indus Entrepreneurs (TiE), for example,
is a not-for-profit network of over 10,000 entrepreneurs, executives
and professionals with roots in the Indus region. TiE has 44 chapters
in nine countries and claims to have access to one of the largest pool
of intellectual capital in world, and to represent economic wealth creation with an estimated value of $200 billion. Similar associations have
also been created by, for example, Chinese and Taiwanese engineers
and software specialists.

Dossani's survey (2002) of Chinese-, Taiwanese- and Indian-born immigrants in Silicon Valley provides interesting insights into the benefits
derived by immigrants from their ethnic linkages. The survey targeted
10,000 members of Silicon Valley's ethnic Chinese and Indian associations and generated 2,272 responses. The survey respondents
shared many features: they were highly educated, entrepreneurial and
interconnected by dense formal and informal networks. Most immigrants held PhD, Masters or MBA degrees. Almost 78 per cent of the
Chinese and 55 per cent of the Indians had stayed on after completing their education in the USA. There were some differences in
occupational status (Figure 4.5). Taiwanese and Indians advanced relatively rapidly up the corporate ladder and mainly occupied executive
posts, while the mainland Chinese mainly held technical jobs. Indians
probably relied more on their English-language skills, and were more
likely to have MBA degrees, while the Chinese were more likely to have
science and engineering educations. Indians were also better able to
exploit inter-company transfers than Chinese.

Diaspora networks were important sources of employment and
investment (Table 4.2). An overwhelming majority of ethnic Chinese
and Indians were employed in firms that had at least two founders
from the country of birth. These firms usually maintained business ties
with the countries of the founder's origin. A surprisingly high share of
respondents also looked to their countries of birth as places to return to
eventually. Over 40 per cent said they would return home for a full-time

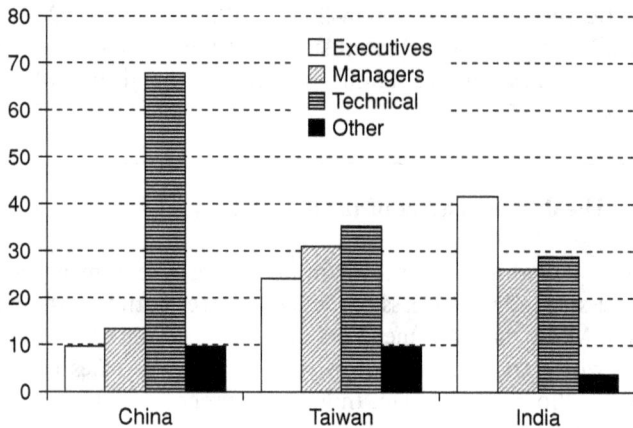

Figure 4.5 Job status of Chinese and Indian immigrants in Silicon Valley

Table 4.2 Transnational activities by Chinese and Indian immigrants in Silicon Valley

	China	Taiwan	India
Number of firms' founders by country of birth (%)			
None	17.5	20.0	11.0
One	31.6	28.0	34.4
Two or more	50.9	52.0	54.6
Number of employers who do business in country of birth (%)			
	60.7	55.1	51.7
Number of respondents who have invested in start-up in home country (%)			
	11.0	15.9	22.3
Likelihood of return to the home country (%)			
To work full-time	42.8	24.8	45.1
To locate a business	78.3	55.4	76.1

Source: Dossani (2002)

job and over three-quarters would consider relocating their businesses there. This trend was particularly pronounced for the younger immigrants. Bangalore in India, Taipei in Taiwan, and Shanghai in China were the preferred destinations. Actual business linkages were weaker than desired, but important. But while most respondents had developed investment activities closer to their current place of residence, a significant share of Chinese and Indians had also invested in their 'home' countries.

Source: after Dossani (2002)

In summary, it has long been evident that national territorial boundaries no longer delineate the extent of the nation's stocks of knowledge. The DKN is a key resources for those 'old countries' that wish to draw on their diaspora, accepting that return migration will be limited. The challenge is the need to create the dense, multiple interconnections that link it and the diaspora (Mahroum *et al.* 2006: 36–7). As Cohen (1997) argues, globalization has not eliminated diasporas, rather it has demonstrated their adaptability (Cohen 1997: xii). 'In the age of globalization, unexpected people turn up in the most unexpected places. Their more diverse geographical spread creates a more truly global basis for the evolution of diasporic networks' (Cohen 1997: 162), and therefore for knowledge transfer.

Transnationalism and knowledge

Transnationalism is a concept that is often linked to diasporas. However, although it may involve ties with an 'old country' this is not necessarily so; for example, the German banker who travels regularly between work in London and New York can be said to lead a transnational life. As with diasporas, the term has many different meanings. For Portes *et al.* (2001: 3):

> 'Transnational fields' is the term coined in the immigration literature to refer to the web of contacts created by immigrants and their home country counterparts who engage in a pattern of repeated back-and-forth movements across national borders in search of economic advantage and political voice.

Transnationalism is considered to have become of significance relatively recently, because only with the advent of modern transport and communications has it been possible for such regular contacts to become normalized in the lives of individuals, or even of entire communities. In some cases, families or partners may lead non-synchronized transnational lives, so that their households 'live together apart' (Hardill 2004).

The contacts that are maintained between countries in transnational lives may be mainly or even purely economic in nature but they can also be cultural and political, and these characterizations may change over time. Although their precise form varies, they are often constituted of people who live dual lives, speaking two or more languages, possibly having homes in two or more countries, and being involved in constant contacts and movements across the boundaries of national spaces (Portes *et al.* 1999). Of course, individuals' lives are transnationalized to varying degrees. In this respect, Guarnizo (2000) identifies what is termed 'core transnationalism' which is understood to involve activities that form an integral part of everyday life, are regular and are patterned and therefore predictable. This is contrasted with expanded transnationalism, which is characterized by occasional involvement in transnational practices. Both forms of transnationalism are important as identifying distinctive means of knowledge transfer.

Transnationalism exists at different scales. Faist (2000) refers to the concept of transnational spaces that consist

> of combinations of ties and their contents, positions in networks and organizations, and networks of organizations that can be found in at least two geographically and internationally distinct places. Not necessarily a definite migration and or return, but instead a circulation, so we can speak of transnational lives.

Another important concept is that of the transnational community (Vertovec 2002), defined by Portes (1997: 812) in terms of 'dense networks across political borders created by immigrants in their quest for economic advancement and social recognition'. The fact that individuals live transnational lives does not necessarily mean that the outcome will be transnational communities, because the latter require the existence of mutual trust and shared identities. Faist (2004) provides an illustration of the linking of transnational communities focused mainly on economic issues, in the form of Turks in Germany and their contacts with Turkey. Initially, these were centred on remittances, although inevitably there was also some exchange of knowledge. Then in the second stage, there was the inception and growth of immigrant businesses in Germany, partly drawing on their knowledge of Turkey for markets and suppliers. Then in the third stage, there was the emergence of transnational production, distribution and sales, which are sustained and also sustain transnational activities, including the constant renewal and transfer of knowledge in and between Germany and Turkey.

There are two main lessons to be drawn from this case study in relation to knowledge transfer. First that transnational communities and individual transnational lives are constantly changing and do not conform to any simplistic model. And second, they are not about 'out there' lives that exist in some discrete transnational space. Instead, as Smith (2005: 237) argues in the context of transnational urbanism, the essence of transnational interconnectivity is that it captures a 'sense of distanciated yet situated possibilities for constituting and reconstituting social relations'. It is this reconstitution of distanciated yet (place) situated relations which characterizes knowledge transactions in transnational spaces. Zhou and Tseng (2002) similarly argue that the networks which constitute transnational activities are only effective if they are firmly anchored in particular locales.

In fact the networks that emerge are not just transnational, but are usually trans-local. Prominent examples exist such as the networks linking Silicon Valley with the Hsinchu region of Taiwan and Bangalore and Hyderabad in India (Saxenian 1999). Transnational organizations like the Monte Jade Science and Technology Association promote business cooperation and technological transfers between the Valley and Hsinchu, while the alumni associations from the Indian Institutes of Technology or facilities like Siliconindia.com play significant bridging roles between Californian and Indian sites (Vertovec 2002: 12). These examples are not drawn at random. Those who lead transnational lives are likely to be skilled or highly skilled, if only because of the costs involved in maintaining these. Zweig and Changgui (1995) found that almost one third of Chinese scientists working in the USA maintained frequent contacts with their home institutions in China. Saxenian (1999) similarly found that Indian and Chinese scientists had strong professional relationships with universities and research bodies in India and China.

In conclusion, then, those individuals who lead transnational lives, either individually or as parts of transnational communities, are in unique positions with regard to knowledge transfer. They learn and transfer knowledge in a situation that involves both international and local connections, and the uniqueness lies in the ways in which they engage in these double-sided knowledge transactions. Highly-skilled workers who lead such transnational lives are particularly likely to have access to uncommon knowledge and to be potentially innovative. However, not all transnational migrants (e.g. elderly German retirees in Mallorca) are engaged in significant knowledge transfers for economic purposes, and they are also not all necessarily highly-skilled workers.

Conclusion: changing mobilities, and changing channels of knowledge transfer

While absolute levels of international migration are at an unprecedented level in the early twenty-first century, in relative terms it is likely that mobility levels are not substantially different from those evident in the late nineteenth and early twentieth centuries. These changes have many drivers, including economic, political and cultural/life style considerations. There are distinctive macro regional patterns of mobility, and these have been shifting over time, from the Americas to Europe to Asia. The changing spatialities are worked out not only at the national level, but also at the regional level with global cities being significant foci in the world map of mobility – although not always in easily predictable ways.

Perhaps as important as the increased scale of migration is the change in the nature of mobility. There have been shifts away from 'permanent' migration and settlement, to temporary, return, serial and circular migrations. There have also been shifts in labour migration towards the mobility of the more skilled – although this has to be seen against significant unregulated labour migration, as well as refugee movements. These all constitute different types of channels for knowledge transfer and learning for individual migrants, with important implications for territorial economies. Even though there has been a shift from economic to cultural and life style motivations among labour migrants, these types of migration flows also have significant tacit knowledge and learning content. We therefore have to think of human capital (and knowledge) transfers not only in terms of transfers from sending to receiving countries, but also in terms of further transfers back to countries of origin, and to third countries in context of serial migration. The economic value of these knowledge transfers depends on the nature of the knowledge, as well as how knowledge transfers are socially situated. In this context, diasporas and transnationalism represent distinctive forms of international connectivity, involving migration and mobility, where the individuals' practices and identities are played out, in different ways, across international borders between two or more countries. This produces opportunities to link and create new forms of knowledge, which increasingly have been recognized by policy makers. In the next chapter, we consider the national and regional levels as key sites in which policy makers seek to leverage migrant knowledge in the interests of particular territorial economies.

5 National and regional perspectives

Innovation and human capital: alternative national strategies

Several research studies have confirmed that the expansion of the knowledge base of an economy can make a substantial contribution to output and growth (Barro and Sala-I-Martin 1995; Soto 2002; Driouchi *et al.* 2006). They have also established that productivity and competitiveness are, by and large, a function of knowledge generation and application. Taking a global perspective, the gap between rich and poor nations is in large part accelerating because of this, but knowledge-intensity can also lead to a growing gap within societies and national economies, whether in social or territorial terms. There is ample statistical evidence that the most advanced OECD economies are essentially knowledge-based, while the least developed nations are not (World Bank 1999). A well-known comparison of Ghana and South Korea is particularly instructive. In 1960 both countries had the same income per capita; today Korea's is seven times higher. The accumulation of physical and human capital explains approximately a threefold difference in income per capita, while the input of knowledge is considered to be responsible for most of the remaining difference.

Lundvall (1992: 8) also made a direct link between economic development and the transfer of (tacit) knowledge. He considered knowledge to be the most fundamental resource in the modern economy, learning the most important process, and interactive learning the most important form of learning. Learning is not confined only to particular specialized nodes, such as universities and R&D units, but is also generated via interactions among institutions within the national system of innovations. These, and similar, views informed economic policies in the 1990s, and were a precursor in some ways to the policy focus on the knowledge-based economy. The OECD (1996), for example, defined knowledge-based economies as being directly based on the production, distribution and use of knowledge and information.

In their approach to strengthening the human capital and knowledge bases of national economies, states face a fundamental decision: whether to invest in developing indigenous human capital and knowledge (via collective expenditure on schools, universities etc.), or to 'free ride' and 'import' these via migration (Straubhaar 2000: 17). The approach taken has varied over time and among national states.

Historically, most advanced capitalist economies fostered indigenous human capital, with varying degrees of effectiveness. In Europe, immigration was encouraged in the 1950s–1970s, but the focus was on relatively 'unskilled' workers to fill gaps in metropolitan labour markets, and there was little explicit reference to knowledge. From the 1990s, there was a shift to more active and positive management of skilled labour migration (Mahroum 2001). In practice, however, immigration policies remain polarized. Skilled immigration is facilitated alongside tighter regulation of unskilled migration, with the latter resulting in increases in irregular migration/asylum seeking (Jordan and Düvell 2002: 2). The first set of policies explicitly recognizes migrants' role in tacit knowledge transactions, while the second effectively denies this.

The dichotomy faced by states is often presented as being whether to invest in indigenous resources for the longer term, or to rely on international migration. The latter is usually portrayed as a short-term remedy but in some cases, for example the USA, consistently high rates of immigration, and conversion of migration to settlement, mean that this is more akin to a medium-term or even a long-term response. In practice, most states seek 'an appropriate balance' (determined as much by political as by economic considerations) between indigenous and migrant labour, while recognizing that they can be complementary as well as substitutes. Policies also have to take into account the rapidity and the duration of adjustments required in human capital and knowledge stocks. Berset and Crevoisier (2006: 79–80) argue that

> In fact, companies are to an increasing extent seeking rare competencies that they can mobilise rapidly. They regularly exert pressure to obtain the necessary work permits. What should be the attitude of the state? Should it allow immigration or finance new types of training for nationals? Different elements must be taken into account: the types of competency required, the durability of the demand, the capacity of the companies to pay salaries for qualified nationals, etc.

In this chapter we consider the role of international migration at the territorial scale, in context of the 'global wars for talent'. First we consider the national level, where immigration policies are mostly framed, and then proceed to consider the regional level where the potential contribution of international migration to learning, knowledge transfers and creativity is effectively articulated.

The national state, innovation systems and migration policies

Although Gamlen (2005) argues that national territorial boundaries no longer delineate the extent of a nation's human capital, the national scale remains an important site for international migration. As Smith (2005: 237) argues, the national state mediates migration because it is 'a repository of language, national cultures and state-centred projects'. Here we focus on three national state 'projects' that shape migration-knowledge transactions: national innovation systems, immigration policies and employment regulation.

National innovation systems

The system of innovation approach was initially developed to study how differences in national organization and institutions can help explain variations in R&D performance and innovative capacities (Nelson 1993, Lundvall 1992). The nature of national innovation systems is contested (Hollingsworth 2000), but the key elements are relationships among firms and other organizations, including universities, public sector agencies and financial bodies, as well as the general regulatory and institutional framework (Lundvall 1992). Debates about such systems pay scant attention to migration, but these shape migrants' knowledge transactions, while migrants can play significant roles in national innovation systems in relation to stocks and flows of knowledge and human capital.

Innovation systems are necessarily complex and are notoriously difficult to manage (Jonkers 2004). They have to address both structures and institutions. In terms of structures, the key features are ownership and scale; for example, the importance of the R&D undertaken in their 'home' countries, or in other centres of research excellence, by transnational corporations such as Nokia in Finland, and Ericsson in Sweden. Such R&D centres are poles of attraction that generate the migration of talent. The broader institutional and regulatory frameworks are particularly related to the educational system, the taxation system and, of course, migration policies.

The key issue for governments is how to upgrade national innovation systems in order to enhance competitiveness. One approach is to borrow institutions and policies that are considered to have been successful in other national settings (Edquist and Johnson 1997: 56–7). Since these borrowed institutions are not introduced into a vacuum, direct institutional imitation is impossible, and successful borrowing depends on careful adaptation to nationally specific institutional settings.

While international migration is often absent from discussions about national innovation systems, the same strictures apply to policy imitation in this arena. The contribution of international migration is probably most obviously observed in the role of foreign scientists in research and development. This has been recognized as a driving force behind the strong R&D performance of the USA (Regets 2001). But there is a two-way relationship, because leading researchers, academics and students are attracted, above all, by key features of the national innovation system including: a critical mass of knowledge excellence; possibilities to work with leading researchers; access to state of the art equipment; and meritocratic and transparent career and reward systems (NOP Business/Institute for Employment Studies 2002). Migrants are also attracted to work for individual firms, or a group of firms, in a knowledge community (see Henry and Pinch 2000 on the UK motor sport industry), and by the openness of cultures to external ideas, or their cosmopolitanism. Therefore, national innovation systems are 'folded' into the regional and firm levels, where many aspects of the infrastructures and cultures of knowledge are articulated. These ideas are further explored later in the chapter, but in the next section we consider the most obvious articulation of migration as a component of national innovation systems, that is immigration policies.

Immigration policies: opening or closing the door to knowledge transfer?

Immigration policies shape the migration channels that migrants utilize – whether illegal or legal, informal or formal – and these also shape their employment prospects (Nagel 2005) and conditions for learning and knowledge transfer. For example, whether individuals are employed in the informal sector, or have been found a job by an employment agency, or are being relocated by a parent company mediates individual knowledge transactions. Of course, the effects of immigration policies are not autonomous of other policy domains (Sassen 2000b), even if the fragmentation of state powers often leads to migration policies being framed with little consideration of, say, employment, housing or training policies. Instead, debates about immigration policies are often highly politicized. Employers tend to favour policies that manage immigration in accordance with their needs – whether in terms of the availability, cost or skills of labour in the economy. National governments, in contrast, find themselves at the intersection of a range of politicized interests (Jordan *et al.* 2003) including those of the resident workforce. This is a difficult balancing act and much of the policy debate has polarized around 'skilled' or 'highly-skilled' versus 'unskilled' migration.

Highly-skilled migration policies have become increasingly prominent because, as Mahroum (1999: 189) argues, 'Nations increasingly view technology transfer as primarily a people-oriented phenomenon. . . . Immigration is thus becoming increasingly an inseparable segment of national technology policies'. This is given additional impetus by the ageing demographic profiles of the more developed countries, and the constrained supply of young, skilled indigenous workers. The resulting general relaxation of barriers to skilled labour migration in the developed world has been well documented (McLaughlan and Salt 2002).

The changing emphasis on skilled labour migration in the developed world, has been part of the shift from restrictive immigration policies to 'managed' immigration. As part of this, 'the skills that migrants bring with them increasingly determine the likelihood of their being admitted in receiving countries' (United Nations 2004a: 78). By the start of the first decade of the twenty-first century more than 20 countries already had policies that favoured skilled labour migration (United Nations 2003). The global talent wars between national states has been expressed in terms of entry conditions, or the precise status granted to skilled migrant workers. Abella (2006: 11–12) comments that: 'In this growing competition for the best and the brightest, traditional countries of immigration which offer permanent residence or at least a path to permanent settlement clearly have an advantage'.

The OECD Science, Technology and Industry Scoreboard (OECD 2006) provides useful data on the international mobility of highly-skilled workers in general, and on students, researchers and scientists in particular. These demonstrate that migration streams are primarily directed towards four economically and technologically advanced destinations: the United States (with over 7.8 million highly-skilled expatriates), the EU (4.7 million), Canada (2 million) and Australia (1.4 million). Migration patterns were determined by both economic and socio-cultural considerations. France, Portugal, Spain and the UK benefited from their colonial histories, while the UK, Canada, Australia and New Zealand benefited from the role of English as the first-order world language (see Alarcon 1999 on Indian migration to the USA). In contrast, migration to, and from, Japan or Korea was limited.

The main destination countries have a variety of immigration policies although there is some convergence through policy imitation. This includes relaxing restrictions not only on the migrant but also his or her family, and providing incentives such as tax relief or employment rights for spouses. In the USA, for example, significant numbers of highly-skilled migrants obtain renewable three-year (H1-B) visas (United Nations 2004a); the quota for these visas was set at 195,000 in 2001, with IT workers from India being one of the main beneficiaries (Martin and Midgley 2003). Several countries also give preferential visa access to foreign students after graduation; for example, this accounts for about half of all the visas granted under Australia's skilled migration programme. European examples include Germany's 2000 Green Card scheme for IT specialists, and the UK's decision in 2007 to move to a points-based system that favours technical skills. In Canada, Quebec offers a five-year tax holiday for academics in the field of health sciences. Despite policy convergence, there are still striking national differences in the openness of immigration policies, even for more highly-skilled migrants.

Abella (2006: 18–19) summarizes the broad approaches to skilled migration among the more developed countries, with most countries combining these in different ways:

- The human capital approach aims to enrich national stocks of human capital over the long term, and typically offers permanent residence and the prospects of citizenship eventually (e.g. Canada).
- The labour market approach aims to address shorter-term skills shortages via a system of temporary visas for migrant workers (e.g. widely used in Asian countries).

- The business incentives approach focuses on investors, entrepreneurs and senior managers, in order to facilitate foreign investment and trade. This usually involves temporary visas but permanent settlement may be offered as an incentive.
- The academic-gate approach aims to attract talented students and then facilitate staying on after graduation (e.g. USA).

The specific policies developed within these approaches are usually framed in general terms rather than being country specific, although they may indirectly discriminate against some national or social groups. Box 5.1 illustrates some of the different approaches taken to recruiting migrant workers in three countries: the UK, USA and Singapore. In all these cases, the destination countries rely on the investments made in human capital by the countries of origin, which raises issues about brain drain and brain gain and welfare distribution (see Chapter 2).

Box 5.1 Diverse international migration regimes: UK, USA and Singapore

UK

The UK has moved from a position of relatively balanced brain exchange flows with other countries, to being a major net beneficiary of brain gain. Whereas migrants from developed countries tend to be temporary, albeit of several years duration, those from less developed countries are more likely to settle or become long-term residents.

The three largest groups of immigrants have been students, asylum seekers and work-permit holders. Between 1999 and 2003, the numbers of student entrants to the UK increased from 272,000 to 319,000, while the number of asylum seekers increased from 110,000 to 185,000. However, the highest growth rates were in the number of work-permit holders, which increased from only some 33,000 in 1994 to over 135,000 in 2003–4. The leading occupational groups were IT, health, and managers/administrators, while the largest national group were from India.

USA

The USA has three main gateways for foreigners: as immigrants, non-immigrants, and unauthorized foreigners. Non-immigrants include spouses, students and temporary migrants who can subsequently apply for immigrant visas: indeed, some 90 per cent of immigrant visas are given to those who are already in the USA. Given that there is a ceiling on immigrant visas for employment, increasing numbers of foreign workers have been using the other two channels. In 2003, there were an estimated 706,000 legal immigrants, and over 1 million apprehensions of unauthorized migrants.

There are several routes to obtain immigrant status but the most substantial is family unification. Some two-thirds of immigrants have family ties to either US citizens or legal immigrants already in the country. Only about one-seventh of the total are admitted on the grounds of having high-level skills, or because employers applied for visas for them. The USA therefore uses a combination of supply and demand criteria for admitting migrant workers.

There are several visa regimes in operation including O-1 visas for migrants with 'extraordinary ability in the sciences, arts, education, business or athletics', and H-1B visas for professional to stay for up to six years. There are also special visa regimes for professionals from NAFTA member countries, Mexico and Canada, L1 visas for foreign nationals being transferred to the USA in intra-firm moves, and for students.

Singapore

Rapid economic growth has made Singapore a major attraction for international migrants, with official numbers rising from about 80,000 in 1980 to 612,000 in 2002, or 29 per cent of the labour force. This far exceeds the proportion in most developed countries.

The largest group of non-resident workers are work-permit holders, undertaking 'unskilled' or service sector jobs in industries facing labour market shortages. However, there are also an estimated 120,000 highly-skilled foreign workers, and three-quarters of these possess employment passes: the eligibility criteria for these are formal graduate or professional qualifications or specialist skills, and wages above the minimum level of S$2500. 'P' and 'S' employment passes, which are determined by skill and salary levels, offer differential rights. In addition, there are EntrePass visas for entrepreneurs and innovators. There are also some 35,000 international students – about 20 per cent of the total student body at state-funded universities.

Unskilled workers have to obtain work permits and these exclude bringing dependents into the country. These are restricted to particular countries and sectors, and employers have to pay a per capita levy for such employees. In contrast, Singapore has one of the most open regimes in the world for employers to hire skilled workers.

Source: after Findlay (2006); Fong (2006); Martin (2006)

In contrast to 'skilled' migrants, 'unskilled' migrants increasingly face tighter immigration rules, often against a background of popular political antagonism. However, the developed countries require unskilled as well as skilled migrants, and indeed the two flows are related, as noted by Sassen (2000a) in the case of global cities. The response, unsurprisingly, has been an upward trend in undocumented migration, whether effected by individuals or as part of migrant trafficking.

Although many countries have enacted measures restricting the entry of undocumented migrants, others have largely ignored the issue. This happens most commonly in sectors experiencing labour shortages, such as agriculture and hospitality services, where both skills requirements and wages are relatively low (United Nations 2004a: 84). Other responses include the use of au pair, working holiday and student visas (allowing part-time working). These schemes tend to offer limited and temporary rights to the migrant and his/ her spouse, and – we contend – pay scant regard to their knowledge potential.

In practice, individual migrants may move between different visa regimes, whether between or within the 'skilled' and 'unskilled categories', as well as from documented to undocumented status and vice versa (discussed further in the next section). As Ruhs and Anderson (2006: 3) state:

> the boundaries between categories are both more permeable and more vague than is often imagined. Immigration status is not static, and an individual's or group's status may change. A group's status may alter when the state changes sets of categories, or the laws or rules governing those categories, or moves people within existing categories.

Despite the shift in favour of skilled immigrants, there is still a persistent 'bricolage of territories with differentiated rights for different migrant groups' (Williams 2001: 103) that has important implications for knowledge transfer. First, there are variations in the extent to which immigration policies specifically recognize knowledge. Policies mostly focus on skills which, given their intangible nature, means relying on qualifications. This implicitly recognizes technical rather than social skills (Reich 1991) and many forms of tacit knowledge, intuition, and emancipatory knowledge. Second, the recognition of knowledge is highly racialized and gendered, as particular jobs and countries are favoured – explicitly or implicitly – in many immigration regimes (see Iredale 2001 on the UK health sector). In contrast, unskilled migrants are mostly valued for meeting shortages of low-cost labour in particular sectors, such as hospitality and agriculture; implicitly they are considered as 'knowledge bereft', and this is reflected in the short-term (often seasonal) nature of the visa regimes that are open to them.

The employment experiences, and potential role in knowledge transfer, of both skilled and unskilled migrants, is determined not only by immigration laws but also by employment regulations.

Employment regulations: more doors to open

Employment regulations also mediate utilization of migrants' knowledge and skills. As with immigration policies, knowledge is usually assessed in employment regulations via the surrogate measure of educational qualifications. Obstacles to knowledge transactions are particularly strong where country-specific qualifications apply to particular jobs, for example in medicine, law and accountancy. However, these qualifications do not necessarily measure substantive knowledge differences: country specific skills are socially and politically constructed, and can be used to protect indigenous workers against competition from immigrant workers or, as Duvander (2001: 210–11) argues, 'to conceal intolerance for diversity'.

Both the national state and a range of intermediate agencies, notably national professional associations, manage employment regulations. Health services have probably been most exhaustively researched. In the UK, for example, the accreditation of foreign qualifications by corporatist professional bodies, such as the Royal Colleges, has mediated the ability of migrants to obtain particular types of employment (Kofman and Raghuram 2005); this has implications for the recognition and utilization, as well as the acquisition, of knowledge. Recruitment practices also tend to channel immigrant nurses from particular countries into temporary and lower-order nursing posts (Hardill and MacDonald 2000), effectively constraining knowledge transfer opportunities. One consequence is differentiation among immigrants: those educated in the host, as opposed to the origin, country will have comparative advantages in the recognition of country-specific knowledge, in terms of their diplomas or degrees. However, there are occupations where (embodied and embrained) tacit knowledge are difficult to represent in surrogate measures such as qualifications. For example, football is one sector where the national origin of the skills acquisition is not important, but where until recently (at least in Europe) the employment regime was highly restrictive for international migrants (Box 5.2). Opera singers and chefs provide other examples.

Employment regulations do not only shape employment and knowledge transactions for higher-order professional migrants. For example, there is relatively limited employment of unregistered migrants in Scandinavia's tightly-regulated labour markets, compared to their substantial irregular employment in 'unskilled' jobs in, for example, catering and hospitality in the UK (Hjarno 2003). These migrants face not only divergent employment conditions, but also different opportunities to use and acquire knowledge, in contrasting regulatory settings.

The combination of migrant status and employment regulation means that individuals face a complex set of opportunities for regulated vs unregulated work under different sets of visa and employment rules. Ruhs and Anderson (2006: 9–10) consider that much of the discussion about 'illegality' or 'undocumented' has confused these different rights, as well as the rights of residence versus the right to work. Moreover, they point out that some migrants may work in the formal sector, paying taxes, but may not be entitled to work at all, or not in that sector, or not for more than a minimum number of hours. They propose an alternative classification based on whether migrants are employment and/or residentially compliant:

Compliance – in full compliance with residence and employment conditions attached to their immigration status.
Non-compliance – not having rights to remain in the UK.

Box 5.2 **Foreign football players, employment regulations and the European Court**

There are two methods of recruitment in professional soccer, namely the transfer and trainee (apprenticeship) systems. International transfers in Europe (and elsewhere) have long been constrained by quotas on the numbers of foreign players, and the fees demanded in inter-club transfers. In 1990, a Belgian football player Jean-Marc Bosman, wanted to move from FC Liège to Dunkirk. When FC Liège tried to stop this move, referring to rules that clubs had to agree a fee before a transfer, Bosman filed a suit against FC Liège, the Belgian football authorities, and the European football authorities. In 1995 the European Court of Justice ruled in favour of Bosman, deciding that existing football transfer rules breached EU law on the free movement of workers between member states (Neatrour and Williams 2002). The ruling benefited the top football clubs and top players in particular. The international migration of footballers in Europe has accelerated sharply since the Bosman case. The establishment of the Champions League, with attendant high levels of economic rewards, and a surge in media

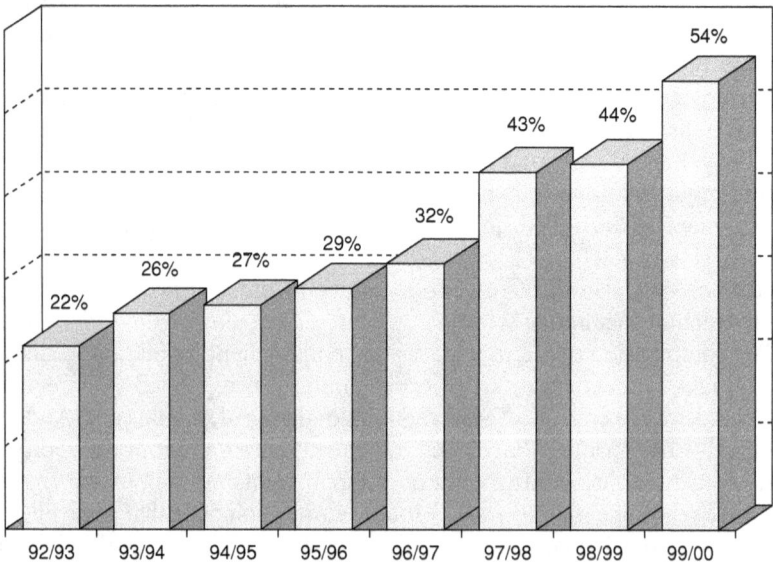

Figure 5.1 The percentage of foreign players in the English Premier League at the start of the season, 1992–2000
Source: after Neatrour and Williams (2002)

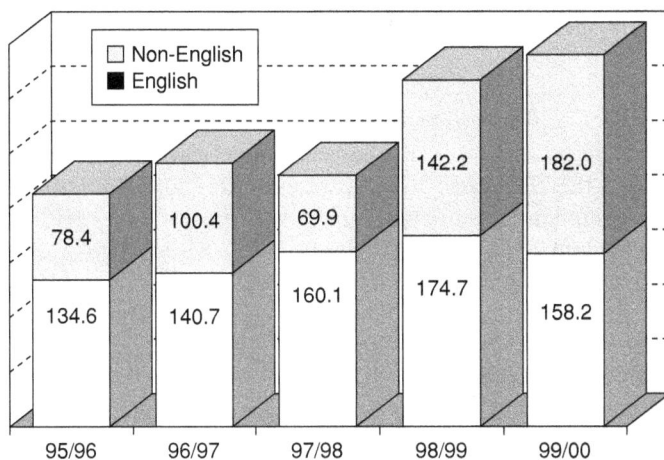

Figure 5.2 Value of transfers between English clubs in the Premier League since the Bosman case (£m)
Source: after Neatrour and Williams (2002)

and sponsorship revenue, also stimulated the internationalization of the skilled labour market for football (Magee and Sugden 2002). By the late 1990s, foreign players outnumbered domestic ones and accounted for most of the transfer value between English clubs in the FA Premier League (Figures 5.1 and 5.2).

While the recruitment of professional footballers has become more internationalized, this trend developed along macro-regional rather than global lines in England. Magee and Sugden (2002) note that English clubs tend to draw heavily on external sources that are broadly similar in terms of culture, language and style of football (for example, Scotland, Ireland, Australia and northern Europe, especially Scandinavia). Large investments in top players are risky because there is considerable uncertainty about the transferability of individual performance. Hiring decisions in professional football are, therefore, strongly influenced by both social criteria and the skills of footballers. High recruitment costs also lead to an uneven distribution of international players as only the leading clubs are able to buy the best players.

Source: Magee and Sugden (2002); McGovern (2002); Neatrour and Williams (2002)

Semi-compliance – have rights to remain in the UK, but working in breach of employment conditions attached to their visas.

Ruhs and Anderson (2006) consider that semi-compliance is relatively widespread in the UK, and other countries, reflecting the need for flexibility in individual lives, as well as in the economy. Of the 576 individuals they interviewed in the UK, some 16 per cent were non-compliant and 7 per cent were semi-compliant. Semi-compliance was particularly evident among those with student and au pair visas.

For individual migrants, the attraction of a country is determined not only by its immigration and employment regimes but also by various other policies that affect their rights, and those of their dependants or spouses. These include their eventual access to citizenship, levels of discrimination, and housing and political rights. The Migrant Integration Policy Index provides a broad guide to this wider policy framework among the EU25 countries, Norway, Canada and Switzerland (Box 5.3).

Box 5.3 The Migration Integration Policy Index, 2007: labour market access

The Migrant Integration Policy Index reviews the application of six policy areas to non-EU migrants in the EU25, Norway, Canada and Switzerland: labour market access, family reunion, long-term residence, political participation, access to nationality and anti-discrimination. The policies are scored against normative standards of best and worst practice. Here we consider only labour market access as illustrating the broad area of employment regulation.

The best case: each practice found in at least one country. A migrant worker or entrepreneur is eligible for the same opportunities as EU nationals to work in most sectors. He/she can count on help from labour market integration measures to adjust to the language and professional demands of the labour market. The state helps the individual to get his/her full set of skills and talents recognized, to access training, and to develop language skills that are critical for the job market. He/she can renew most types of work permits and remain living in the country and look for work if they lose their job. With job security come equal rights for all workers. He/she is free to change employer, job, industry and workpermit category in pursuit of professional development, and also has the right to join a trade union.

The worst case: each practice found in at least one country. In an excluding labour market, a migrant's skills and qualifications from

---

country of origin are not recognized or are downgraded through an unfair, long and costly procedure. Even if qualified, the migrant is not eligible to work or become self-employed in many sectors, due to government restrictions. Without access to any labour market integration measures, training, or study grants, it is difficult for the individual to overcome language and professional barriers. Status as a worker is insecure. Rigid administrative criteria prohibit retention of a work permit if fired. Even if a company wants to keep the migrant, these criteria prohibit renewal of the permit. Migrants wishing to stay securely in the country are bound to particular employers because they are denied the right to change employer, job, or industry.

The 28 countries in the study are scored and ranked in relation to the six policy areas, but here we present only the grouped rankings for labour market access.

Table 5.1 Index of migrants' labour market access

Score	Countries
91–100	Sweden
81–90	Italy, Portugal, Spain
71–80	Belgium, Canada, Estonia, Switzerland
61–70	Finland, Netherlands, Norway
51–60	Lithuania, Slovakia, Slovenia, UK
41–50	Austria, Czech Rep., France, Germany, Ireland, Luxembourg
31–40	Cyprus, Denmark, Greece, Hungary
21–30	Malta, Poland
11–20	Latvia

Source: http://www.integrationindex.eu/topics/2585.html, accessed 31 October 2007

Academic and scientific recruitment: migration at the heart of the national innovation system

The attraction of academic and scientific workers is the most obvious way of upgrading a national innovation system via international migration. Not surprisingly, the more developed countries have attracted not only large numbers of the academic and scientific talent of less developed countries, but also the most able members of these groups. A recent estimate suggested that some 400,000 scientists and engineers from developing countries (between 30 per cent and 50 per cent of the total stock) work in research and development in the industrial countries, compared with 1.2 million doing similar jobs in their countries of origin (Meyer and Brown 1999). Jamaica is an extreme case: there were nearly four times more Jamaicans with tertiary education in the US than in Jamaica itself in 2000 (ILO 2004).

The most skilled academic and scientific workers are particularly mobile: 12 per cent of Mexico's labour force, but 30 per cent of its PhDs are in the USA. Moreover, doctoral students are especially likely to stay on in the USA after completing their studies: 87 per cent for students from China, and 82 per cent from India (OECD 2002). Some countries' doctoral programmes have a strong foreign student presence. Non-citizens account for more than a fifth of all doctorates in Switzerland, the UK, the USA and Australia. The UK attracts PhD students from most European countries, while Spain and Portugal attract students from Central and South America, reflecting linguistic and cultural ties. Asian students are the largest element among foreign doctorates in the USA (OECD 2006).

The European Union has gone furthest in attempting to remove (internal) barriers to academic and scientific mobility. Other than the general freedom of movement provisions, there has been an attempt to promote the idea of the European Research Area, and gradual harmonization of higher education qualifications. The notion of the European Research Area was quite simply to strengthen the research capacity of the EU via enhanced mobility among other measures. Removal of formal barriers to mobility is only part of the picture, for there are still considerable national differences in employment and working conditions for academic and scientific staff among European countries. Scientists and academics are, of course, attracted by higher salaries but they are also attracted by employment and working conditions – they value independence, good facilities, and working in centres of excellence. Fair and transparent recruitment and promotion practices are also highly valued (Ackers 2004). This is borne out by a study of the motivations of highly-skilled migrants to the UK (Home Office/DTI 2002). They particularly stressed career advancement, global centres of excellence, and the quality of research facilities, in addition to salary differentials. Chompalov *et al.* (2002) consider that high levels of mobility are facilitated in scientific labour markets because these are comparatively small, and the work of individuals is highly visible in the form of publications and conference performances. Mobility is however highly gendered in the academic and scientific communities as in most labour markets (Ackers 2004).

While we have emphasized aggregate numbers in this section, the impacts of academic and scientific mobility on national innovation systems are shaped not only by quantity but also by quality (Mahroum 1999), that is the ability of countries to attract or retain the most able, and most knowledge-able individuals. This is one reason why most of the developed countries are keen to attract the brightest international students to study at their universities and to try and retain them. Such organized and explicit migrant learning and knowledge transfer is increasingly a feature of many national innovation systems.

Student migration: attracting and retaining potentially knowledgeable workers

Khadria (2001) contends that student migration represents 'semi-finished' human capital migration. This is attractive to destination countries for several reasons. Not only as a source of fee income for universities, but as a way of adding to the pool of potential knowledgeable workers, who have nationally specific human capital, if they can be persuaded to stay on. Not surprisingly, a number of countries have targeted student immigration.

There is considerable scope for student mobility, not least because of the highly uneven actual, or perceived, standards of national higher educational systems. Krieger (2004) reports that some 10 per cent of students in Bulgaria and Romania had firm intentions of emigrating, although we should note that these levels are likely to change rapidly in transition economies. The receiving countries may find that such students have a high propensity to stay on after graduating. Whereas, previously, this was viewed negatively or at least suspiciously by many European countries, they increasingly incentivize staying on. Even where students do return, they usually retain strong links with the country where they studied. Moreover, undergraduates who have studied abroad are more likely to have further periods of research or work abroad. This may be a re-emigration of failure if they cannot find the desired employment and living conditions in their home country, but their student experiences may also have given them a new appetite, awareness and confidence for further migration (Baláž and Williams 2004).

As with scientific and academic mobility, student mobility is highly unevenly distributed. Eighty per cent of all foreign students in the OECD countries are found in just five countries: the USA, UK, Germany, France and Australia (Abella 2006). The gap between the three European countries and the rest of the leading EU countries is considerable in absolute terms, but less so in relative terms (see Box 5.4).

Countries adopt different policies to attract foreign students (Kuptsch 2006). In the late 1990s, Germany set out to 'internationalize studies in Germany' by promoting Germany as a centre for science and education. In the UK, the government announced a target of attracting 75,000 additional students by 2005, by streamlining visas, enhancing information, and easing restrictions on foreign students taking part-time jobs. This was followed in 2000 by the French government announcing an ambition to double the number of visas issued to foreign students. In practice, the decisions of potential students about where to study is influenced by several considerations, including fee levels and the cost of living, the reputation for academic excellence, the prospects of getting temporary jobs while students and well-paid permanent jobs after graduating, and the visa requirements for family members.

Having attracted students, most countries now also encourage them – particularly in science and technology – to stay on (Abella 2006: 22–3). This

***Box 5.4* Foreign students in Europe**

There is a highly uneven distribution of foreign students (as indicated by citizenship) across Europe, with the largest absolute numbers being in the UK, Germany and France. However, Belgium, Austria and Switzerland are relatively more important if percentage data are considered. The most impressive growth in absolute and percentage terms since the 1990s has been in the UK, which has the advantage of working in the first ranked international language.

Table 5.2 Foreign students in selected European countries: rank order, 2001

	Foreign students	*Share of all students (%)*
UK	225,722	10.9
Germany	199,132	9.6
France	147,402	7.3
Spain	39,944	2.2
Belgium	38,150	10.6
Austria	31,682	12.0
Italy	29,228	1.6
Sweden	26,304	7.3
Switzerland	27,765	17.0
Netherlands	16,589	3.3

Source: after OECD (2004: Table 3.2)

Source: Kuptsch (2006: 34)

takes many forms. For example, facilitating the switch from student visas to work permits as in Australia, Canada, France and Ireland. Loans and subsidies may also be made available to encourage the more able students to complete their graduate studies, and this is in addition to competitive scholarship schemes. In the UK the government launched the Science and Engineering Graduate Scheme (SEGS) in response to shortages of indigenous graduates in some branches of science and mathematics; this allowed students to stay on and seek jobs in the UK for a period after graduating.

The big issue, both for the individuals and for national economies, is whether the students will stay on. Student motivations for migration are initially mixed and may change during the course of their studies. As Li *et al.* (1996) elegantly phrase this, are students 'migrating to learn or learning to migrate'? Staying on rates are variable in practice, both between origin and destination countries, and disciplines. The USA, and to a lesser extent the UK, tend to retain relatively high proportions of foreign students but particularly those from some countries (Ackers 2005). For example, in the USA 88 per cent of Chinese PhDs stayed on for more than five years, compared to only 11 per cent of Korean PhDs. Given that the university years are a critical period in the formation of friendships and family, many students may be induced to stay on not only by career opportunities but also by personal ties (Kuptsch 2006: 41).

Policies to encourage returned migration

The loss of highly-skilled migrants had been viewed as a permanent loss by many countries or origin, but there has been growing interest in the potential to use the knowledge and capital of the diaspora, as well as measures to encourage return migration (see also Chapter 4). A number of countries have introduced specific incentives to encourage return – such as Korea and South Africa. Others have relied more on economic and social transformations to create an environment that is more conducive to return (Lowell and Findlay 2002; OECD 2002).

Zweig (2006) provides a useful guide to the issues surrounding the attraction of return skilled migration. A number of countries have experienced low rates of return among the students and scientists who had gone abroad to study, notably China and India. However, rates of return have increased in recent years, driven by rapid economic development in these and other emerging industrial economies, increased opportunities for enterprise, and – in some cases – encountering a glass ceiling on further advancement in the destination countries. China exemplifies the shift in national policies to encouraging return migration (Box 5.5).

The attraction of such return migration is that: 'Returnees seem to be of higher quality than people who have not gone abroad' (Zweig 2006: 203), whether measured in terms of grants, fellowships or publication. Of course, emigration was selective in the first place, and many of these returnees were among the most able and talented members of their cohort at school or university. So care must be taken not to exaggerate the value-added knowledge acquired by migration. Nevertheless, this does not in any way undermine the potential value of attracting these individuals to return. There is however, another issue and that is whether return is reverse selective, that is whether the most able individuals opt to stay abroad. Zweig (2006) reports a view that the top 20 per cent were not induced to return to China. The evidence on this is problematic but it is notable that a webbased survey of Indian IT professionals in Silicon Valley – a centre of undoubted excellence – found that about half of respondents aged 35 or less did not intend to return to India. There are, then, limits to the extent of return – although this survey does show considerable potential to attract back some leading scientists and technologists (see also Saxenian 2006 and Chapter 7).

This has been particularly impressive in the health sector in India but also in the IT sector, where brain drain has been turned into brain training by increasing levels of return migration (Zweig 2006). Returnee Indian doctor

Box 5.5 **Mother China wants you back**

The decade of the Cultural Revolution (1966–76) decimated the Chinese academic community. Since 1978 China has sent students and scholars abroad in order to rebuild its human resources in science and technology. Chinese leader Deng Xiaoping argued that even if 5 per cent did not return, the policy would be a success. In fact China suffered a severe brain drain with little reverse flow of knowledgeable migrants. Only one-third of the estimated 380,000 students and scholars who had gone overseas since 1978 have returned (Figure 5.3). Since the 1980s, China's central government has tried to lure students back. It offered returnees preferential treatment in terms of better housing, faster promotion and higher social status (Zweig 2006). By the late 1980s these policies were being copied by regional and local governments in the major Chinese cities. The most highly-valued returnees were generously provided with tax incentives, opportunities for start-up business loans, new laboratories and research grants, salary supplements and/or privileged access to quality housing and cars. While these policies mostly failed in attracting back the brightest talents in terms of international reputation and prestige, they nevertheless had some success. Many returnees settled in China's export and high-tech development zones and have

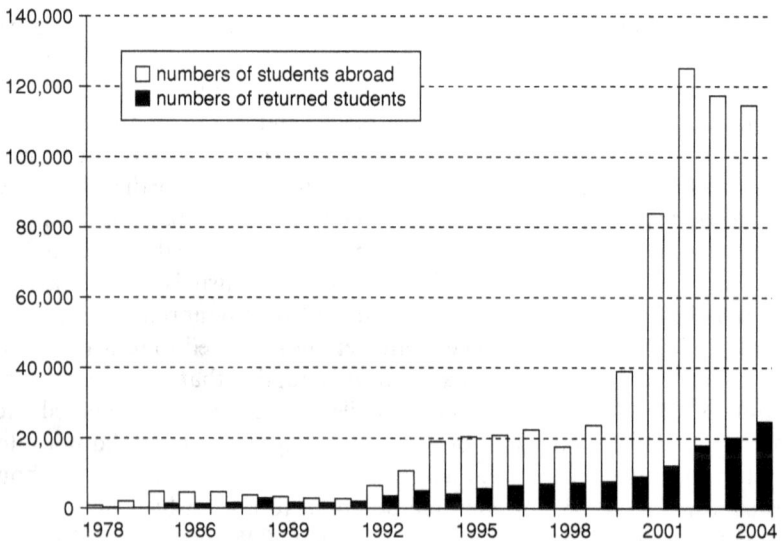

Figure 5.3 Chinese students abroad and returned students, 1978–2004
Source: after NBSC (2006)

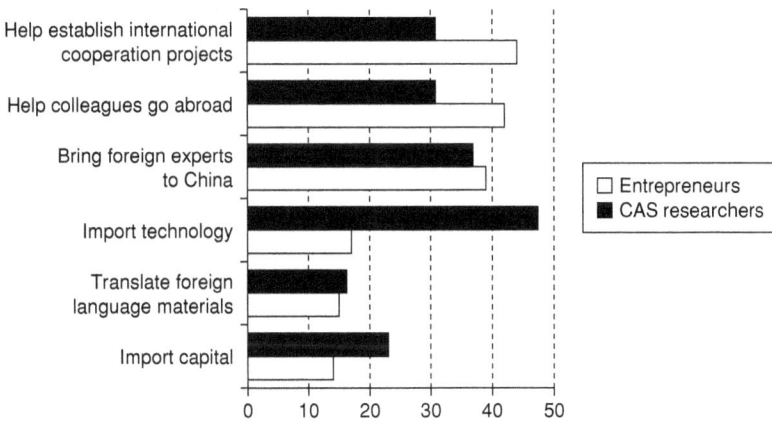

Figure 5.4 Forms of international cooperation: evaluation of relative importance (%)
Source: after Zweig *et al. (2004)*

brought knowledge and technologies that have had a positive impact on the Chinese economy.

Zweig *et al.* (2004) evaluated the importance of various forms of China's access to foreign intellectual and financial resources. Two samples of Chinese returnees were questioned: researchers from the Chinese Academy of Science (CAS) and entrepreneurs developing their businesses in Guangzhou, Beijing and Shanghai. The creation of transnational networks emerged as the most substantial issue in international cooperation, Both groups considered bringing foreign experts to China; helping colleagues go abroad and establishing international cooperation projects were far more important initiatives than inward flows of financial capital (Figure 5.4).

Since 1995 the Chinese government has also tried to exploit the huge potential of 'brain power stored overseas'. It has appealed to Chinese non-returnees, who should 'serve China from abroad'. The overseas students have been invited to invest in China, set-up technology and consultancy businesses and participate in other forms of knowledge and technology transfer. In many respects, China has been replicating proactive human resource policies developed by Hong Kong, Taiwan and South Korea. These took advantage of rapid economic development to reverse brain drain and attract skilled migrants with financial and human capital and transnational networks to return home.

Source: after NBSC (2006); Zweig (2006); Zweig *et al.* (2004)

entrepreneurs from the UK, the US and the Middle East have helped to set up world-class private hospitals and clinics in India. Not only do they have capital, but they also have knowledge and networks. They have helped to transfer the latest equipment and technology to these Indian hospitals. Of particular note is the Apollo group of hospitals, India's first corporate hospital chain, established by Dr Pratap Reddy, who had returned from the USA. Apollo has become one of Asia's largest health-care establishments, attracting significant foreign investment. There is also evidence that the growth of private hospitals in Hyderabad was largely driven by returned migrant doctors, particularly from the Middle East. In addition to their specialist technical knowledge and networks, the doctors also realized other types of knowledge transfer:

> There are other intangible benefits, including knowledge of quality processes, better ethics and attitudes towards work, greater professionalism and transparency, better management practices, and familiarity with the latest technology. Other intangibles include reputation and brand equity benefits as returning doctors also help in building the reputation of the institutions in which they work. The presence of returning super-specialists and overseas trained doctors helps these establishments convey an image of quality care and treatment of international standards to patients.
>
> (Zweig 2006: 238)

The regional scale: institutions, innovation and factor mobility

As emphasized in the Introduction, we have adopted a multi-layered perspective on the role of international migration in knowledge transactions, seeing the regional as one scale embedded in, or folded into (Amin 2002), other scales; in other words they are mutually informing. Whereas the particular significance of the national scale is that this is the key site for the regulation of international migration, and the determination of the national innovation system, the regional scale is particularly significant because it is the scale at which many of the key institutions (collective learning cultures, cosmopolitanism) and infrastructures (whether housing, or airport connections) are constituted and shape migration. Amin and Thrift (1994) expressed the former in terms of regions being a locus of cultural and institutional specificity – and this resonates with the concepts of encultured and embedded knowledge (Blackler 2002). These two dimensions are instrumental in shaping the attractiveness of a region for migrants, labour-market spillovers, and social networking, all of which shape migrants' knowledge transactions. Our approach to defining the regional is deliberately flexible, and incorporates a range of scales between the small town on the one hand, and the national on the other.

If knowledge is assumed to be the main driving force of innovation, and the latter is the major determinant of competitive advantage in a globalizing economy, regions are key sites for considering migration-knowledge relationships. There have been many attempts to conceptualize the driving forces of regional development. Theoretical views on these issues have moved from comparative advantage (based on natural endowments) to competitive advantages (based on intra-industry trade and localized demand conditions for market competitiveness) to constructed advantages. The latter is sometimes described as the 'Triple Helix' model (Etzkowitz and Leydesdorff 2000) and embraces the surplus value of an overlay of relations among the three components of a knowledge-based economy: (1) the knowledge-producing sector (science); (2) the market; and (3) governments. Close interactions among these partners are facilitated via spatial proximity. 'Learning regions' became a key concept of this approach when explaining the relationship between knowledge and economic growth at the regional level (discussed in the following section).

Almost every region in Europe (and in many other countries) has tried to foster an innovative environment based on public or semi-public innovation in infrastructures and private-sector-oriented policy measures. Innovation is seen as the key to rapid economic growth, higher levels of regional wealth and the restructuring of regional economies towards a more sustainable basis – whether in economic, social or environmental terms, or some combination of these. Physical capital and natural endowments are no longer considered the driving factors behind economic growth. New competitive conditions favour social capital, and formal and informal norms and rules, which can foster reciprocal understanding and mutual confidence among the key agents in the regional economy (Cooke and Morgan 1998). Modern regional policies pay considerable attention to identifying and developing infrastructures that are conductive to innovation, to facilitating networking and cooperation among companies (and key centres of knowledge such as universities and public research agencies). There is recognition that to be effective this requires integrated actions by the regional administration, and the support of the most influential agents in the region. The story of Finland, which was transformed in the 1990s from a medium-tech, Soviet market-oriented economy to a high-technology performer, has been widely been admired, generating numerous attempts to emulate it by both national and regional economies. Yet, few countries were able to repeat the successes of Finland and other Nordic countries. In part this is because of the timing of their entry into key new-product markets, such as the mobile phone, but it also underlines the specificity of institutions and the ways in which these interact to shape knowledge transfer and innovation. By the mid-2000s, there were diverse levels of innovation and knowledge inputs in the regions of the enlarged Europe, and these were strongly related to regional income differentials (Box 5.6).

Box 5.6 Innovation regions in Europe

There is evidence to support a positive relation between a region's innovative performance and its economic performance. Empirical data from the European Regional Innovation Scoreboard (RIS) indicate that about 41 per cent of the variation in per capita regional income can be explained by differences in innovative performance. A laggard innovative performance may thus lead to widening or persistent regional income disparities.

European regions display diverse levels of innovativeness. Only a minority of regions engage in highly sophisticated knowledge-based production. They have rich networks of R&D facilities, large stocks of human capital and can effectively commercialize the outputs of investments in R&D. In most EU countries, less than one-third of the regions perform above the country mean, confirming the regional polarization of national innovative capacity. The leading regions in the EU in terms of R&D expenditure and employment in Knowledge Intensive Services (KIS) are Braunschweig, Oberbayern and Stuttgart (DE),

Table 5.3 Top 15 ranked EU regions in terms of GDP per head, R&D expenditure and employment in knowledge-intensive services (KIS)

	GDP per capita in PPS per capita in 2003		Total R&D Expenditure in 2002, as % of GDP		Employment in KIS in 2004, as % of total employment	
1	Inner London (UK)	60 342	Braunschweig (DE)	5.25	Inner London (UK)	59.83
2	Brussels (BE)	51 658	Västsverige (SE)	5.19	Stockholm (SE)	54.74
3	Luxembourg	50 844	Stuttgart (DE)	4.42	Oslo og Akershus (NO)	49.76
4	Hamburg (DE)	40 011	Stockholm (SE))	4.37	Outer London (UK)	49.21
5	Île de France	37 687	Oberbayern (DE)	3.71	Utrecht (NL)	49.08
6	Vienna (AT)	37 158	Strední Cechy (CZ)	3.18	Surrey, East & West Sussex (UK)	49.03
7	Berkshire, Bucks, Oxfordshire (UK)	35 894	Sydsverige (SE)	3.10	Bruxelles-Capitale	48.68
8	Bolzano (IT))	34 792	Tübingen (DE)	3.06	Övre Norrland (SE)	48.65
9	Oberbayern (DE)	34 334	Pohjois-Suomi (FI)	3.06	Zürich (CH)	47.86
10	Stockholm (SE)	34 331	Noord-Brabant (NL)	2.82	Île de France (FR)	46.97
11	Åland (FI)	33 542	Östra Mellansverige (SE)	2.78	Trøndelag (NO)	46.92
12	Utrecht (NL)	33 148	Mittelfranken (DE)	2.62	Berlin (DE)	46.95
13	NE Scotland (UK)	32 683	Etelä-Suomi (FI)	2.60	Berkshire, Bucks, Oxfordshire (UK)	46.65
14	Southern and Eastern (IE)	32 446	Darmstadt (DE)	2.55	Nord-Norge (NO)	46.57
15	Darmstadt (DE)	32 251	Länsi-Suomi (FI)	2.38	Mellersta Norrland (SE)	46.47

Source: based on EU (2006)

Brussels (BE), Stockholm, Västsverige and Sydsverige (SE), Inner and Outer London and Surrey (UK), and Noord-Brabant (NL) (see Table 5.3). In contrast, most of the poorest performers are located in the new EU Member States.

Regions with large shares of employment in Knowledge Intensive Services (KIS) employment and large R&D expenditure also feature in the highest-ranked region in the EU in terms of income. When the evidence from the RIS and R&D are combined, the role of KIS in regional economic development is even clearer. All the RIS indicators show a significant correlation with per capita GDP, with the highest correlation coefficient being for high-tech services employment.

Source: based on EU (2006)

The Community Innovation Survey (CIS) is a particularly useful source of information on knowledge creation and transfer at the firm level and competitiveness. The survey is undertaken jointly by the European Commission and the statistical offices of the member states of the European Economic Area (EEA). The CIS contains a wide range of information, including data on firms' main markets (domestic versus international) and internal and external sources of knowledge (in-house R&D, clients, suppliers, consultancy, trade fairs, universities). Analyses of CIS databases have demonstrated that enterprises with innovation activity tend to focus on national/international markets, while enterprises without innovation activity tend to focus on local/regional markets. Within the EU, almost half (49 per cent) of all enterprises without innovation activity reported that their most significant market was local or regional (irrespective of whether or not it was in the same or a neighbouring country). Among enterprises with innovation activity, less than one-third (30 per cent) reported their most significant market as being local or regional. In contrast, a quarter of enterprises with innovation activity indicated that their most significant market was international, compared to 13 per cent for enterprises without innovation activity. Studies of leading innovation regions in the UK (Simmie 2003), Netherlands (de Bruin 2004), Finland (Thomi and Böhn 2003), Sweden (Nählinder and Hommen 2002) and the USA (Porter 2003) have suggested that these benefited significantly from international trade and knowledge transfers. These regions are much more flexible in responding to technological changes and shifts in demand. Innovation was significantly related to growth in exports of goods and services, and international collaboration in R&D. The suppliers of innovative firms, on the other hand, tended to be of local/regional origin.

Linked to the concept of constructed advantage, we can suggest that there are two basic drivers of regional innovation. First, the international:

globalization, exports, and the movement of production factors (capital and knowledge in particular, including via human mobility). Second, the regional and local: agglomeration effects; cluster effects; and regional policies aimed at innovation.

Simmie (2003) argues that the major differences in the drivers of regional innovations stem from the relative importance of tacit and explicit (transferable) knowledge in regional innovation systems. There are diverse mechanisms for creating, transferring and implementing knowledge at the regional level. These include informal mechanisms, learning by doing, learning by interacting with customers and suppliers, the mobility of personnel and more formal R&D activities. The most common ways of transferring tacit knowledge are social networks and the mobility of labour (including international migration). Examples of tacit knowledge transfer include:

- R&D cooperation among universities, research facilities and businesses;
- movement of R&D personnel;
- interaction between public and private sector (industry-academia relations);
- interaction within the private sector, for example between a firm's customers and/or competitors;
- technology diffusion.

The availability and configuration of these mechanisms translate into the region's capacity to create knowledge via dynamic interplays among key actors. Regional systems of knowledge significantly depend on existing physical infrastructures and socio-economic milieus, which are necessarily differentiated, constituting important factors in the reorganization of economic spaces (Park 2004). Tacit knowledge is considered essential for fostering innovations in general and therefore also for regional innovations. Building social networks requires frequent interactions among the major stakeholders in innovation. These interactions are facilitated by, although not necessarily dependent on, spatial proximity. Some regions have significant agglomeration effects, which are substantial for building knowledge infrastructure. The rich texture of regional innovation infrastructures (universities, R&D facilities, and knowledge centres) is also an important precondition for developing social networks. These networks, in turn, facilitate knowledge transfers and enhance the economic growth and competitiveness of a region.

The two driving forces of regional innovation – internationalization and regional agglomeration and clustering effects – initially appear contradictory. Internationalization and globalization support the rapid spread of explicit knowledge (directly via ICT or indirectly via imported technology), while regional clustering favours transmission of tacit knowledge due to spatial proximity. Knowledge created in more distant regions is more likely to be uncommon and its transfer offers potentially higher returns. Spatial distance, however, limits personal interactions that are often crucial for transferring subtle forms of knowledge. Firms enjoying spatial proximity can benefit from larger numbers of face-to-face interactions and socio-cultural environments.

The former support trust building and decrease the transaction costs of knowledge transfer. Horizontal clusters of firms enable mutual learning via monitoring and competition among its members, Vertical clusters provide for different type of learning based on specialization and cooperation of complementary networks of suppliers and customers. Specialization and physical proximity support economies of scale, decreasing transaction costs and facilitating innovative solutions by specialist firms. Similarity of socio-cultural environments, however, decreases the likelihood of radical differences in knowledge levels. The process of 'knowledge homogenization' is further reinforced where there are high levels of regional industrial specialization. Despite this rich theoretical literature, few studies have actually found evidence of the superiority of local over non-local interactions. Even regions previously held to be models of clustering (for example, Oberbayern and Stuttgart) were in fact highly engaged in international movements of production factors as well as by dense regional interactions. The strong and weak points of intra- versus extra-regional knowledge transactions are presented in Table 5.4 in relation to the role of human mobility.

Table 5.4 Intra- and extra-regional knowledge transactions via human mobility

	Intra-regional	*Extra-regional and international*
Frequency of transfer	High frequency and density of face-to-face contacts. Many opportunities to build trust.	Lower frequency and density of face-to-face contacts.
Social, cultural and economic environments	Social, cultural and economic environments of knowledge brokers and CoPs likely to be similar and do not impinge on knowledge transfer.	Social, cultural and economic environments of knowledge brokers and CoPs may be very different and substantially impact on knowledge transfer.
Costs of knowledge transfer	Low costs of tacit knowledge transfer.	Higher costs of tacit knowledge transfer via personal mobility.
Potential for tacit and explicit knowledge transfer	High potential for transfer of tacit knowledge.	High potential for transfer of explicit knowledge. Transfer of tacit knowledge is more difficult and may by aggravated by 'knowledge noise'.
Knowledge gradient	Relatively low degree of novelty and uniqueness of shared knowledge. More limited choice of alternative knowledge sources (over-embeddedness).	Radically different knowledge types offer greater opportunity for above-standard returns. Introduction of evaluation and control mechanisms for knowledge implementation may be costly.

Source: authors

The next section explores the role of international migration in these knowledge transactions.

Migrants and localized knowledge: pipelines and buzz

Berset and Crevoisier (2006) consider that migrations play an essential role in maintaining competitiveness in regional production systems (RPSs) in two different ways. First, via the attractiveness of regions for the mobility and localization of production factors, including labour. They consider this to be a quantitative perspective, dealing with how labour responds to wage or employment differences between different economic spaces. In other words, much of the focus is on neo-classical equilibrium models. Second, in terms of regional transformation and innovation:

> These RPSs do not maintain their advantage thanks to lower costs, but rather thanks to their capacity for making their specialised competencies evolve dynamically and autonomously (following their own logic) as a result of integrating new knowledge. In such a perspective, migrations are not seen as correcting an imbalance between a source and target region. Migration makes it possible to *maintain the competitiveness and coherence of an RPS*.
>
> (Berset and Crevoisier 2006: 63; emphasis in original)

In this perspective, the key to understanding the contribution of migration lies in the nature of the circulation of people and their competences between regions, and the process of 'anchoring the workers in the firms that employ them' (Tarrius 1996: 2001). Flynn's (1985) study of the region of Lowell in the USA provides an example of how, what Berset and Crevoisier (2006) term 'the circulation of competencies', via inter-regional migration, has contributed to technical change in a traditional industrial economic base – initially favourable, but later problematic in the case of Lowell.

A more holistic view of knowledge in this context is provided by the learning region concept. The literature on learning regions (having a high level capacity, via innovation, to adjust to changing economic conditions) starts from the assumption that tacit knowledge is most effectively transferred, face-to-face, by those who share similarities in terms of language, social norms, and personal knowledge developed through long-established formal and informal interactions. A key contention is that physical proximity facilitates trust, which in turn facilitates knowledge transfers and collective learning (Maskell and Malmberg 1999).

Subsequent writings on this topic have acknowledged the importance of knowledge sourced from outside, as well as within the region. There has been an extensive critique of the tendency to essentialize the role of proximity in knowledge transfer. Oinas (2000) argued that proximity only facilitates interactions, and does not necessarily create them. And, while distance may hinder interactions, it does not exclude them. Moreover, the assumed relationship between trust and proximity has been questioned (Hess 2004: 175). Amin (2002) similarly argues against privileging spatial proximity because firms draw on a variety of networks, at different scales ranging from the local to the international. He argues that physical proximity and localized face-to-face contacts are not essential for developing trust-based relationships: instead, 'intimacy may be achieved through the frequent and regular contacts enabled by the distanciated networks of communication and travel' (Amin 2002: 393–4).

Recent commentaries have emphasized that territorial knowledge clusters are a multiplicity of sites, characterized by both local and distanciated connections (Amin and Thrift 2002: 51–2). Clusters do have localized advantages, what Storper and Venables (2002) termed the urbanization economies of 'buzz' (dense possibilities for face-to-face knowledge exchanges). However, the most successful firms or regions seeking, to benefit from knowledge creation and transfer, exploit both intra-regional and extra-regional sources of knowledge. Bathelt *et al.* (2004: 38) discuss this in terms of 'buzz' and 'pipelines'.

- The co-presence and co-location of people and firms within the same industry, and place or region, promote face-to-face contacts and generate local information and communication ecology, or *'buzz'*. This '. . . consists of specific information and continuous updates of this information, intended and unanticipated learning processes in organized and accidental meetings'.
- Radically different knowledge is more likely to be sourced externally – that is from outside the shared knowledge of the region. *'Pipelines'* or channels for tapping far-off knowledge resources are built through strategic partnerships at the inter-regional and/or international scales. This may be realized electronically or via corporeal or human mobility. Knowledge transfer via 'pipelines' is more difficult to accomplish and has relatively higher transaction costs than 'buzz', not least due to differential financial and time costs of travel. Overcoming socio-cultural differences and a lack of shared knowledge (embedded or encultured) can also add to the transaction costs of externally sourced knowledge. However, 'pipeline' costs can be transformed by the introduction of new communication and transport technologies (see Box 5.7).

International migration is one way in which knowledge can be transferred between regions. Most of the earlier writings on 'learning regions and cities' did not ascribe a role for migration, and it was left to Maskell and Malmberg (1999: 18) to make explicit an underlying negative assumption: 'If . . . mobility is sufficiently low, the owners and managers of firms in most industries in a region or a small country will know each other either directly or indirectly'. In other words, migrants lack the encultured and embedded knowledge to be effective participants in these knowledge transactions.

***Box 5.7* Low-cost airlines and knowledge transfer**

There have been several studies of the impacts of ICT on the diffusion of explicit knowledge. Far less attention has been paid to how technological and organizational changes impact on the transfer of tacit knowledge. Economic spaces are constituted as rich networks of institutions, which both shape the activities of market participants and are shaped by these. Although economic agents (firms, organizations, networks etc.) are subject to economic, technological, institutional, cultural and other constraints, they are also creative forces that can shape these constraints. Radical innovations and shifts in techno-economic paradigms created by the interactive learning activities of innovative agents may lead to adaptations in institutional frameworks. The emergence of low-cost airlines (LCA) and the distributed networks of labour force recruitment (for example, via the internet) may be considered as a co-evolution of technical and institutional change.

The growth of LCAs can generate both quantitative and qualitative (structural) changes in labour markets. Due to decreasing travel costs and increasing numbers of travel destinations, there is what Harvey (1990) terms 'time–space compression'. This has consequences not only for the frequency of travel, but also for the social distribution of travel. Reductions in the financial and time costs of travel mean new social groups are willing and/or able to travel further distances (including migration) to secure jobs. Workers who would not be prepared to travel by bus, or to be away for many months, might be willing to take the same extra-regional jobs if they could travel by air and/or commute weekly. In addition, LCAs offer other advantages to firms. For example, a KPMG (2005) survey in Hungary found that a growing number of businesses have opened offices at, or relocated their businesses close to, the airport from which low-cost airlines fly. High-frequency and -density air travel effectively redefines borders between intra- and inter-regional knowledge transactions. LCAs can potentially change geographical proximity and the spatial reach of 'buzz' knowledge conditions.

What are the consequences of these developments for knowledge transfer? LCAs impact on knowledge transfer in a number of ways:

- First, LCAs directly contribute to knowledge transfer. Tacit knowledge, as we have noted (Chapter 3), is bound to its bearers. There are various channels through which the knowledge exchanges can take place, namely labour mobility, inter-firm links, communities of practice, knowledge communities, market and semi-market institutions and networks. All these channels are based on human interaction. There is a debate as to the relative importance of local compared to 'distanciated' relationships in the transfer of knowledge, as noted in this chapter, but there is overall consensus that some degree of human mobility is critical to tacit knowledge transfers, even if partly substituted for by electronic transfers. In this case LCAs have an important contribution to make: the more people travel, the more tacit knowledge potentially can be transferred.

- Second, LCAs have indirect impacts on the development of new markets and technologies. LCAs may lead to increased knowledge about technology, for example via personal visits by managers and professionals to other firms in other regions, or to exhibitions. If firms have the capacity to acquire and use this technology, it may lead to either: (a) increased and more efficient use of existing technology; or (b) purchase of new technology. In addition, we can assume that firms which predominantly have foreign customers are more innovative than those having domestic ones. To access foreign markets, firms need to have specialist knowledge, which is not widely available in the domestic economy.

- Third, LCAs also contribute to organizational flexibility (the evaluation of locational advantage, or changes in the spatial organization of the supply chain). In general, LCAs do not carry large volumes of cargo, but they could be important: (i) for industries with low-volume, high-value inputs and outputs; (ii) for moving spare components quickly between plants (whether intra- or inter-firm suppliers); (iii) for being able to send engineers and other technicians quickly and cheaply to remedy local problems at extra-regional branches.

- Fourth, LCAs may, via migration by highly-skilled workers, contribute both to reproduction or removal of disparities in regional stocks of knowledge. Labour migration is likely to have different consequences for the countries of origin and destination. The country of origin loses labour, which mostly tends to be at least middle-skilled. In the short term, this contributes to brain drain. Most migrants, however, learn new skills. As many eventually return, their country of origin benefits, in some instances, from brain gain in the long term. Whether the brain drain / brain gain results in a zero-sum game or win-win depends on the conditions of knowledge transfer between migrant countries of origin and destination.

- Fifth, the regional distribution of the impact of LCAs on international migration and knowledge transfer depends on the channels of migration. Skilled migrants tend to depend on professional agencies/companies to facilitate their mobility – so their distribution is likely to be geographically relatively dispersed as a result of LCAs providing more direct city-to-city connections. But unskilled workers tend to rely more on social networks of friends and family: therefore, chain migration tends to concentrate them in areas of existing immigration, that is, the metropolitan areas.

Source: after Williams and Baláž (2008a)

We can think of knowledge transfers via migration in terms of knowledge flows. The flows are shaped by what Urry (2000: 35) terms scapes, which are 'networks of machines, technologies, organizations, texts and actors that constitute various interconnected nodes along which the flows can be relayed'. These flows become well-travelled routes linking particular places, such as the migrations from Portugal to France in the 1970s, or from Romania to Italy in the 2000s. They are sustained by particular transport technologies, networks and organizations that facilitate particular flows, as opposed to others. Bunnell and Coe (2001) suggest that the 'astronauts' shuttling between Taiwan's Hsinchu region and Silicon Valley are iconic figures for mobility (Saxenian 1999), but they can also be thought of as travelling along a scape for highly-knowledgeable migrants. Other Silicon Valley examples include transnational migrants who originated in India, China or Israel, and are now strongly embedded in both locations (Saxenian and Hsu 2001). Of course, not all migrants travel along these strongly-articulated scapes, especially in the case of learning regions, where the search for highly-skilled migrants is more likely to be global. However, the global is articulated through particular networks, and connections between particular places – in other words, for migrants it is a highly-structured space made up of interconnected places.

Linked to this notion of interconnected places, Bathelt *et al.* (2004) argued that external knowledge acquired through 'pipelines' can overspill to other local firms via 'buzz, or through generalized labour turnover (Keeble *et al.* 1998: 20). A number of UK case studies (Henry and Pinch 2000; Smith and Waters 2005) empirically demonstrate the importance of high rates of localized labour market turnover in knowledge spillovers, but largely ignore links between migration and such spillovers. Yet, in-migration is generally strongest in dynamic urban regions and these, of course, are also likely to be areas with high labour market turnover. Moreover, migrants may be attracted to regions with routinized high levels of labour turnover, being indicative of high levels of employment opportunities and reduced labour market precariousness. However, the attractiveness of regions to individual migrants is determined by more than intra- and inter-firm mobilities – it also depends on infrastructure, cosmopolitanism vs localism (see next section) as well as differential wages and employment conditions.

The relationship between international migration and collective learning and knowledge transactions is most highly developed in iconic high-tech areas. One of the few studies to address this relationship is Alarcon's (1999) study of Silicon Valley (but see also Saxenian 1999). Immigrant workers played a key role in innovation: 'a flexible industrial system in a knowledge based industry depends on a flexible immigration system that allows the continuous arrival of workers with the newest skills and ideas' (Alarcon 1999: 1395). Foreign-born engineers and scientists in Silicon Valley account for nearly one-third of the engineers, mathematicians, computer scientists and natural scientists employed in the high-technology industry in Silicon Valley. This is a considerably higher proportion than in the Route 128 cluster in New England where the corresponding figure is only 17 per cent. This difference seems to be due to two factors: the existence of a larger immigrant pool in Silicon Valley and the openness of the industrial system to new migrants. Interviews with engineers from India and Mexico, who had worked in both regions, confirmed the more open environment in Silicon (Alarcon 1999: 1385). While Silicon Valley is probably exceptional, it has informed other national and regional projects to assemble critical masses of 'idea generation' (Lucas 2001), via recruitment of highly-skilled-scientists and engineers, many of whom are migrants.

Finally we note that the literature on knowledge communities also pays limited attention to the role of migration in knowledge transfer, but does at least recognize this implicitly. Henry and Pinch (2000) analyzed the agglomeration economies of the British motor sport industry, concentrated in 'Motor Sport Valley', and identified key elements in its constitution as a knowledge community. These include labour market features, such as rapid and continual staff transfers within the industry, the convergence of careers (most skilled workers spend at least part of their careers in this particular cluster), labour market churning due to the deaths and births of firms, and non-labour market factors. The constantly shifting pool of skilled labour within and from outside the knowledge community, that is including (circulating) migrants, is of particular relevance.

In summary, migration does intersect with collective learning in learning regions or knowledge communities. The learning regions and knowledge communities perspectives provide frameworks for understanding the circulation of migrants' translated knowledge from an initial employer to the wider labour force in a particular territory. In other words they address the issue of how knowledge is anchored within regions. However, both approaches are unduly economistic, and myopic to a degree, failing to address how discrimination and stereotyping shape opportunities and constraints for migrants (see Chapter 6 for a discussion of these issues at the firm level).

Creativity and cosmopolitanism

Migrant learning and knowledge transfers are socially situated, that is they are not only confined to the workplace but are distributed across work *and* non-work places. Of particular note here is the way in which migrant experiences are both shaped by the general (extra-workplace) socio-cultural environment, and contribute to this in terms of creativity, cosmopolitanism and social diversity. Labour migrants have diverse motivations, including economic, cultural and lifestyle objectives. Many labour migrants have narrowly economic objectives, including acquiring qualifications or skills (and associated knowledge), which have economic value in either the destination or the origin region. For them, engagement with the culture of the destination region is a necessary cost rather than a benefit (Hannerz 1996: 106). However, other migrants value 'openness' to cultural diversity and cosmopolitanism. As noted earlier, this partly explains the greater relative importance of foreign-born engineers and scientists in Silicon Valley compared to Route 128, two major high-tech clusters in the USA (Alarcon 1999).

Creativity is a key element in innovation: 'Innovations are the practical application of creative ideas, and an organization cannot innovate unless it has the capacity to generate creative ideas' (Westwood and Low 2003: 236). It is fed by cultural diversity but there are differences in how the latter is interpreted. Sassen (1998: xxxi) recognized the social and cultural diversity of large western cities, but also emphasized that for many less-skilled migrants corporate power inscribed many cultures and identities with 'otherness'. Her view was grounded in the persistent structural relationships that dominate major cities. Florida (2005: 36) was more positive – but also partial in ignoring many forms of less-skilled migration – and argued that the 'creative class', key players in knowledge creation and transfer, sought 'diversity of all kinds'. Migrants are attracted by, but also contribute to reproducing, diversity as a defining feature of dynamic and creative city regions. And diversity facilitates creativity and openness to knowledge transfers from migrants. In reality these are not so much competing views as indicative of selective readings of particular types of migration and different social settings.

Social and cultural diversity also has a negotiated set of meanings for individuals. It is very different for young professionals in the cultural industries compared to internationally mobile managers and professional in, say, the financial and business services, let alone for seasonal agricultural workers or kitchen staff. Cultural diversity is likely to be integral to those working in the culture industries, attracting them to particular places – often cities, where migration has played a key role in shaping a cosmopolitan culture. In contrast, there is evidence that managers and professionals from major companies are more likely to be tightly focused in their social and professional networking, and inhabit relatively smaller sub-sets of the cultural diversity of the city, especially if their mobility is shorter term (Beaverstock 2004; 2005). Other migrants may end up in jobs in the enclave economy, and may also become relatively isolated from the wider society if, additionally, they live in ethnic enclave communities. Their experiences of cultural diversity may be truncated at best, and – if they encounter discrimination and prejudice – may be very negative. In the remainder of this chapter, we focus on two positive interpretations, Richard Florida's notion of the creative class and creative cities, and the cosmopolitanism of (some) global cities.

Richard Florida, the creative class and migration

Florida (2002) argues that, in the knowledge economy, territorial competitive advantage is based on the ability to mobilize rapidly a combination of skilled people, resources and innovations. This capacity stems from being able to generate, attract and retain an effective combination of talent, creative people in the arts and cultural industries, and diverse ethnic, racial and lifestyle groups. His views are reinforced by Lee *et al.* (2004) who stress the need for creative people, from varied backgrounds, to come together to generate knowledge and innovation. For them diversity is a type of social infrastructure and this is a reserve that entrepreneurs (and we could add, all workers) can draw on. One of the features of diversity is relatively low entry barriers that facilitate migrants and others being able to utilize their human capital (knowledge) in a particular region. Lee *et al.* (2004: 882) argue that immigrants are more likely to be entrepreneurs, but that in the USA there are two very different groups: the highly educated versus the unskilled who are often reluctant entrepreneurs because of the constraints they faced as migrant employees. Both are risk takers but the first are more likely to feed off and feed the creative city.

Global cities are key nodes of creativity: Amin and Thrift (2002: 59) consider them to be meeting places for knowledgeable and creative people, and sites of knowledge transfer. Similarly, for Jacobs (1961) cities are 'open systems' which attract talented people from diverse social and geographical backgrounds. Of course, international migration does not play an equally prominent role in all global cities (see Box 4.2), and Tokyo and Seoul for example lack this migration-based social diversity, especially in relation to skilled workers in the cultural industries.

The role of migration in shaping and sustaining social and cultural diversity is addressed by Florida (2002: 750–1). A, and sometimes the, key challenge for firms is how to produce and retain talent because 'high human-capital people have many employment options and change jobs relatively frequently'. They seek locations where there are multiple job opportunities, while the 'creative class' in particular seek out high-quality experiences, openness to diversity of all kinds, and opportunities 'to validate their identities as creative people' (Florida 2005: 36). Therefore, the importance of social identities and cosmopolitanism for knowledge transfer and creation is significant both within firms (see Chapter 6) and within particular places.

While providing useful insights, these theories understate the discrimination and stereotyping faced by migrants, which shape the opportunities and constraints they encounter. For example, Zhou and Tseng (2002: 142) show that many Chinese high-tech entrepreneurs in California followed this pathway because they encountered a glass ceiling as employees. Theories of creativity also make sweeping assumptions about the openness and cosmopolitanism encountered by migrants in their non-working lives. In short, there is need for a more non-essentialist understanding of territorial competitive advantage, and the role played by migrants in this. The final section of this chapter considers issues relating to cosmopolitanism.

Global cities, cosmopolitanism and migration

Hannerz (1996: 131) argues that individuals, in what may broadly be termed the cultural industries, tend to migrate to global cities when they are relatively young. This is partly because these provide unique learning opportunities, but also because of a desire to be 'in the right place'. Cosmopolitan individuals want to be able to immerse themselves in other cultures: 'they want to be able to sneak backstage rather than being confined to the front stage area' (Hannerz 1996: 104). These global cities are 'open systems' (Jacobs 1961) that attract internal and international migrants from diverse backgrounds. Similarly, Beaverstock (2002: 525) comments that as transnational elites 'flow into or through the city, they bring with them well-established cosmopolitan networks, cultural practices and social relations'. Hannerz (1996: 129–98) also notes the relationship between transnationalism and cosmopolitanism, writing that transnational relationships 'play major parts in the making of contemporary world cities'. Individuals with transnational lives can move relatively easily between global cities, with only limited personal dislocation costs.

What is meant by cosmopolitanism? There are competing definitions, of course, but Hannerz (1996: 103) is particularly helpful:

> cosmopolitanism can be a matter of competence. . . . There is the aspect of state of readiness, a personal ability to make one's way into other cultures, through listening, looking, intuiting, and reflecting. And there is cultural competence in the stricter sense of the term, a built-up skill in manouvering more or less expertly with a particular system of meanings (i.e. general and specific). . . . Competence with regard to alien cultures itself entails a sense of mastery, as an aspect of the self. One's understandings have expanded, a little more of the world is somehow under control.

Cosmopolitanism is easier to theorize than to measure. Nevertheless, it is clear that there are considerable differences between countries in this respect. For example, a survey of 450 major organizations in the larger EU states explored the diversity of their senior management (Dassù and Franklin 2005). On the one hand, in Italy only one of the top 75 organizations is led by a foreigner, in France there are just two foreign leaders among the top 75 organizations, and there are only three or four in Sweden and Spain. In contrast, there was a far more international picture in Germany and the UK. In Germany, foreigners head 12 out of 75 organizations, although many are Austrians or Swiss Germans, while foreigners in the UK lead 26 organizations, although 17 are from other English-speaking countries. There is then evidence of significant differences, in terms of this simple index, even though language mediates the way that cosmopolitanism is played out. It can also be noted that only one company, even in the UK, had a chief executive who was a member of an ethnic minority: Arun Sarin, an Indian-born US citizen, was the chief executive of Vodafone.

Most migrants are not of course cosmopolitan – no more than travelling automatically turns an individual into a cosmopolitan. Cosmopolitanism means being open, and while travel may make some individuals more open to encounters with diverse cultures, for others it may produce a negative and inward-looking perspective. The limits of individual cosmopolitanism are defined not only by the individual but also by his or her relations with others. The ability to express and practice cosmopolitanism depends in part on reciprocal relationships with others. It is evident therefore that it is significantly easier to practise cosmopolitanism in some places than others. Global cities are particularly fruitful places for cosmopolitan individuals to engage in knowledge transactions with like-minded individuals, especially where open-mindedness becomes contagious. At the other extreme are low-learning and inward-looking regions, where diversity and creativity are likely to be relatively low, and it is difficult for individual migrants to be beacons of knowledge transactions.

Conclusions

Our starting points in this chapter are that innovation is a key driver of modern economies, that human capital is a key element in innovation, and that migration is a significant component of the latter. Migration is particularly important in innovation because this requires the application of tacit and codified knowledge and, as we have argued throughout this book, migration plays a particular role in the transfer of tacit knowledge. Moreover, as the discussion of cosmopolitanism and creativity have indicated, the contribution of international migration to the social diversity that underpin these is often critical.

The brief review in this chapter has confirmed that, as would be expected, there are highly uneven distributions of innovation at different scales. This reflects both structural and institutional conditions at both the regional and national scales. Theoretical arguments can be constructed in favour of the importance of endogenous versus external knowledge in national and regional innovation systems, but there is an emerging consensus that what matters is how these intersect, or are combined. These issues are being carried over into policy debates, particularly at the regional level. At the national level, the critical policy areas of immigration and employment regulation focus more on skills than specifically on knowledge and innovation. There is a constant global scan among policy makers who are seeking models and best practices, but these are embedded in local or national institutional frameworks, which represent opportunities and obstacles to their transfer between countries and regions.

Human mobility can be a critical factor in innovation development and transfer of tacit knowledge, if the appropriate policies are applied at the national, regional and firm levels. State policies are important in determining the overall framework for channelling human capital into national economies, but they are only a precondition. The regional and firm levels are also important in a multi-layered approach. This chapter focused particularly on spatial proximity versus migration, and on creativity and cosmopolitanism at the regional level. The next chapter addresses the firm level.

6 Firm-level perspectives

Knowledge, mobility and the firm

Leveraging learning and knowledge transfer is a, if not the, key to the competitiveness of the firm or the efficiency of organizations in the management literature (Gues 1997; Nonaka and Takeuchi 1995). Nonaka *et al.* (2000) criticize 'traditional' concepts of the firm developed by some western scholars as being static. Neo-classical theory views a firm as an equilibrating mechanism, where inputs of production factors are combined in order to achieve maximal profits. Profit maximization is based on rationality and perfect information on the operating environment, and leads to a stable point in the long run after a series of adjustments. In practice, it is recognized that this long-run point may not be reached because of constant changes in the operating environment. In contrast, transaction costs theory and principal agents theory recognize bounded rationality and the opportunism of economic agents but, like neo-classical theory, view the firm as a relatively static entity, which processes information taken from the external environment in order to solve specific problems and adapt to the environment. Nonaka *et al.* (2000) criticize such views as static. They argue that firms not only solve problems, but also create and define them and, in this way, generate knowledge through problem solving.

We can therefore see firms as 'repositories of competences, knowledge, and creativity, as sites of invention, innovation and learning' (Amin and Cohendet 2004: 2). Our particular perspective in this volume is of the firm as an entity with blurred boundaries. There are systems of knowledge within the firm, but these are linked to systems of knowledge outside the firm and, as we have seen in the previous chapter, these are multi-scalar.

Nonaka *et al.* (2000) provide a useful starting point for conceptualizing these relationships and knowledge 'systems'. In the knowledge-based theory of the firm, a firm is viewed as a community dealing with knowledge creation and transfer. Nonaka *et al.* (2000: 8) introduce the concept of 'ba' (a Japanese word for 'place'), which they consider to be a social community, where knowledge is shared and created. It does not necessarily have to be a physical place, but a specific, fluid time space created through interactions among individuals and/or individuals and their environment. This notion parallels ideas about encultured and embedded knowledge (Blackler *et al.* 1998; see Chapter 2) but with greater emphasis on the role of 'community'.

In the view of Nonaka *et al.* (2000), the firm is a dynamic configuration of 'ba', whose boundaries are both open and closed. Knowledge in the firm is created via spiral-like interactions between tacit and explicit types of knowledge. The so-called SECI spiral consists of four modes of knowledge conversion.

(1) socialization (from tacit to tacit knowledge);
(2) externalization (from tacit to explicit knowledge);
(3) combination (from explicit to explicit knowledge); and
(4) internalization (from explicit to tacit knowledge).

Nonaka also presented the concept of knowledge assets, which are seen as inputs and outputs of knowledge-creating processes. A distinction is made between experiential and routine knowledge assets, which are tacit in nature, while conceptual and systemic assets are explicit in nature. These knowledge assets determine the firm's boundary – as many of them may lie outside the formal or legal borders of the firm, for example in communities of practice or suppliers. They also determine the costs of the inputs for knowledge-creation processes. At the same time, such processes determine the accumulation of the knowledge assets of a firm. A firm's boundary is thus not determined via legal ownership of particular assets (e.g. land) or contractual 'ownership' of workers, but by a firm's ability to build and energize dynamic 'ba'.

What are the keys to building and energizing such knowledge communities? As Nonaka *et al.* (2000) suggest, the key is the relationship between tacit and explicit knowledge and the linking of different types of knowledge assets. The nature of knowledge transfer to, from or within an organization depends on the share of the tacit component in that knowledge. Knowledge with a greater degree of tacitness tends to be more valuable, but also more difficult to transfer. Where does human mobility fit into this? It is of course an important vehicle for knowledge transfer as we have emphasized throughout this book. The knowledge transfers may be planned and contained within the borders of the firm, or unplanned overspills. Migrant workers can play different types of roles in knowledge transfers, and this is captured by Barley and Kunda (2004) in their terminology of 'gurus, hired guns, and warm bodies'. Gurus are bought in to provide knowledge that permanent employees lack, while warm bodies and hired guns provide just-in-time knowledge over short periods. Each of these types of migrant-based knowledge transfer can be effected within companies, although organizations also rely on externally sourced labour. We return to intra-firm moves in the next section of this chapter. Here we focus on the extra-firm movements of migrant workers. While they represent knowledge gains for the recipients they may represent losses for the originating firms: these may be absolute (where the knowledge is irreplaceable) or relative (allowing competitors to reduce the originator's knowledge advantage). This is one reason why firms may pay a premium to retain highly knowledgeable workers (Box 6.1).

***Box 6.1* Returns to tacit knowledge**

Human capital theory assumes that young workers have greater incentives to invest in training and knowledge building than older workers. The latter have fewer incentives to make such investments because they have fewer economically active years ahead in which to secure returns on their accumulated human capital. The theory also suggests that young workers in intellectually demanding professions are more ready to accept below-market pay in the initial stages of their employment. They take a longer-term view of returns, and consider that building up a stock of valuable knowledge will pay off in the later stages of their careers.

Employers will be ready to pay above-market wages to employees with high stocks of valuable knowledge, in order to prevent knowledge spillovers via the movement of workers to their competitors. The incentive for employers to pay higher salaries is particularly associated with non-explicit types of knowledge, which cannot be transferred other than via labour mobility.

Møen (2005) found some evidence for these assumptions. He analyzed the development of experience and the wages of technical staff in the machinery and equipment industries in Norway, 1986–95. These industries were chosen for their absolute sizes and high R&D intensities. The sample was based on annual data for some 30,000 workers in 750 plants. Regression analysis confirmed that technical staff (workers with secondary technical education and scientists and engineers) received below-market wages in R&D intensive firms in the early and middle stages of employment, but were rewarded with significant wage premiums in the later stages of their careers.

- Scientists and engineers working in R&D intensive firms had average wages that were lower by 2.7 per cent in their first year of employment than those of colleagues starting in firms with no R&D. However, scientists and engineers with more than 35 years of experience had average wages 4.4 per cent higher than similarly experienced colleagues in firms with no R&D.
- Similar patterns emerged for workers with secondary technical education in R&D intensive firms. They had wages lower by 4.1 per cent in the early stages of their careers, but higher by 4.7 per cent in the final stages, compared to colleagues in R&D non-intensive firms.

- In total, the employers' current R&D intensity reduced wages for workers with less than 20 years' experience and rewarded those with 20+ years experience.

R&D usually is performed by workers with higher education. Wage premiums for workers with secondary education in R&D intensive firms indicated that these firms were ready to pay not only for direct research exposure, but also for technological training at all levels within the enterprises. Møen comments on the negative effects of R&D on initial pay, arguing that firms' technology levels are more important for wage trajectories than formal on-the-job training. He suggests that while formal training is short-term, working in a technologically challenging environment supports accumulation of human capital throughout the duration of a job.

Møen also investigated the effects of labour mobility on pay rises. Workers with technical or scientific education in R&D-intensive firms who stayed with the same employer had higher wage growth throughout their career than those moving among firms. Above-market wages for loyal employees essentially create barriers to labour mobility and increase firms' ability to appropriate rents from R&D activities.

Source: after Møen (2005)

Knowledge spillovers is a useful concept for developing our understanding of such unintentional knowledge transfers. Arrow (1962: 615), writing about knowledge spillovers, emphasized that 'no amount of legal protection can make a thoroughly appropriable commodity of something so intangible as information'. It is difficult therefore to control the spillover of knowledge across the legal or the effective boundaries of the firm. Mobility of workers, managers, consultants, suppliers and others between firms provides one way whereby such knowledge spillovers are realized.

As Geroski (1995: 78) states, 'spillovers occur when a researcher paid by one firm to generate new knowledge transfers to another firm (or creates a spin-off firm) without compensating his/her former employer for the full inventory of ideas that travels with him or her'. They may travel across the road to a neighbouring firm, or at any scale up to the inter-continental. For example, Song *et al.* (2003: 352) comment on the importance of 'learning-by-hiring', a process that was critical in the growth of major Korean companies such as Samsung, which deliberately recruited scientists and engineers who had worked for market-leading US firms.

Of course, not all workers have equal weight in knowledge spillovers. The focus is on certain key individuals, who have access to 'uncommon' knowledge that can contribute to enhancing value added in the recipient firm. In other words, it is about the match between the migrant's knowledge, and the firm's capacity to absorb and apply that knowledge. As emphasized in Chapter 2, migrants do not carry some unproblematic and specific technical knowledge, for this is also socio-economically and culturally contingent to different degrees. All knowledge is socially and culturally constructed and tacit knowledge in particular has different meanings and applicability in different environments.

Experts do not simply collect knowledge via experience and observation, but also apply heuristic methods to these. They draw analogies in recognizing structural features and in finding a common denominator between two or more issues. Creative and flexible methods enable experts to reorganize such knowledge into complex hierarchical systems, and develop complex networks of causally-related information (Epstein 2005). Migrants face the challenge of moving knowledge not only between workplaces, but also between institutional settings at the national and regional levels. The extent to which these challenges can be overcome determines the potency of the knowledge spillover for both the originating and receptor firm, as well as the originating and receptor economy. We return to the management issues that this presents for firms in the final section of the chapter.

Here we can note more generally that learning and knowledge transfers are influenced by the conditions of migration. Although it is difficult to generalize on this point, we can draw attention to some key issues:

- Whether mobility is bounded (e.g. within intra-company transfers) or unbounded, as part of free-moving boundaryless careers (see Chapter 7). The former usually provide more structured opportunities for co-learning and knowledge transfer.
- The immigration channels utilized (Nagel 2005): were they registered or unregistered, and did they intend to be permanent or temporary migrants? This mediates the types of jobs and industries migrants seek employment in. There is, in short, an intersection of migration regulation, and processes of co-learning and knowledge transfer.
- The nature of the employer organization is important. This is partly an issue of firm size and complexity. Howells (2000: 54), for example, argues that the distance between the knowledge frames of individuals tend to be greater in larger firms, or firms spread across multiple (geographically-diverse) sites, where staff are more likely to be drawn from different cultural backgrounds.

Beyond these generalizations, micro processes within the firm are influential in organizational learning (Andrews and Delahaye 2000) and we consider where migration fits into this picture in the remainder of this chapter.

The transnational company and migration: mobile corporate culture

Salt and Findlay (1989) outlined the need for a new conceptual framework for skilled migration, identifying a threefold typology of mobility: transfers within transnational companies' international contracting of staff to work abroad, and recruitment by international agencies. While the last of these has belatedly been researched (for example, Ward 2004), intra-company mobility has received most attention and is the focus of this section.

Despite recent interest in this topic, the role of different types of mobility, including human migration, in knowledge transactions within transnational companies remains imperfectly understood. The starting point is the challenges facing transnational companies, which have to seek ways of expanding and managing international systems of production and distribution. In essence, they have to balance a company-wide strategy (for example, to integrate its diverse operations) with the need to engage with locally or nationally specific conditions of production (Millar and Salt 2008). This poses the need to balance, and integrate, both locally and distanciated knowledge transactions (Amin 2002).

There are four main reasons why international mobility is important to transnational companies. First, there is a need *to assemble project teams* to work together on highly specific tasks, which may be related to innovations or research and development. Gibbons *et al.* (1994: 6) noted how these operate: 'People come together in temporary work teams and networks that dissolve when a problem is solved or redefined. Members may then reassemble in different groups involving different people, often in different loci, around different problems.' Grabher (2004: 104) adds that although working in project teams is not new in some industries, such as movie production or construction, it has taken off in a broad range of industries in recent years, leading to 'projectification'. Although such project teams usually involve the physical assembling of the teams, perhaps involving international mobility, he also emphasizes that they have distinctive 'project ecologies', that is networks of enduring ties and institutions that persist after the specific project is completed.

Second, transnational companies may need key professional or managerial workers to *establish their operations in new countries*, particularly if there are shortages of such skilled workers in these countries (Erten *et al.* 2006: 42). Millar and Salt (2008), for example, comment on how mergers and acquisitions are often the trigger for mobility. Market development is another important trigger of international mobility, driven by both cost and efficiency considerations. However, they consider that in most of the aerospace and extractive companies they studied, product-, project- and process-related factors were the main driving forces of mobility. The early stages of production are characterized by high levels of costs and uncertainty, and perhaps the development of complex technologies. These may generate substantial mobility flows in the start up period, which decline in the latter stages of the product life cycle.

Third, there is a need among transnational companies *to distribute their collective knowledge*, particularly their business culture, across widely distributed branches. In the earlier human resource literature on this topic, the emphasis was on the use of management transfers as part of companies' strategies to control often-sprawling business empires (Edstrom and Galbraith 1977). This constitutes a means for a headquarters to increase its control over its branches through personalized controls rather than more formal mechanisms such as budget controls.

This view is, however, based on one type of corporate organizational model where power and knowledge are seen as being concentrated in the core or headquarters, and the challenge is how to diffuse this down through the corporate hierarchy and across borders (Edwards 1998). In contrast, Bartlett and Ghosal (1989) identify four different types of internationalized firms – multinational, global, international and transnational – differentiated in terms of their degree of centralization, and linkages across establishments. Human mobility is relatively low in the first two types, whereas core-branch mobility is implicit for knowledge transfer in the third, and genuine trans-company mobility is implicit in the fourth.

Fourth, companies may *value their managers and professional workers obtaining experience of working in different countries*. To varying degrees, internationalized companies value the acquisition of international experiences because they are one articulation of the 'transformational experiences' that force people to engage with new ideas and practices: 'It is now de rigueur for high-potential managers to be given an international assignment' (McCall 1997: 77). Leading transnational companies usually require what Perkins (1997: 62–3) terms an 'international cadre of executives' who are:

> capable of transferring the enterprise's commercial and operational philosophies and systems into every location in which they wish to do business. This group – capable of thinking global, acting local, and vice versa – will be among the premium capital any organization will wish to have access to.

However, this 'rationale goes little beyond an assumption that a "stint overseas" can be quite developmental' (McCall 1997: 77). The 'transformational experience' may be related to the generic experience of working in a different setting (learning to see things differently, gaining self-confidence etc.) or to specific national (or even metropolitan) knowledges: establishing networks with other professionals or with potential clients, or understanding local business cultures in specific territorial settings. Downes and Thomas (1999) emphasize the importance of this function because it is seen as the acquisition of rare, or even unique, (knowledge) resources from which can stem long-term competitive advantage. In fact, what companies value is the acquisition of encultured and embedded knowledge, and an ability to reflect critically upon these. There may also be recognition of the distinctive nature of such knowledge, and the potential for individuals to become boundary spanners (Tushman and Scanlan 1981) (see Chapters 2 and 3).

Despite the lack of clarity as to the content of these experiences, the important role of international mobility is not in question. Several surveys (for example, NOP Business/Institute for Employment Studies 2002) have shown that companies value international migration as a means to foster cultural diversity (via migration) and redistribute international expertise (via migration and return migration) across their branch network. A broadly similar picture is painted in Perkins (1997). Beaverstock (2005), Beaverstock *et al.* (2000) and Beaverstock and Boardwell (2000) also provide evidence of how important such international experiences are in financial services, in cities such as New York, London and Singapore. These studies identify three main reasons why companies in this sector send staff to work abroad: to obtain or transfer specific knowledge within the company, including the perpetuation of organizational culture; networking and accumulating cultural capital; and performing 'global facetime' processes between firms and clients. These broadly correspond to our earlier identification of the distinctiveness of migrant knowledge, and the importance of encultured and embedded knowledge. In addition, transnational experiences are also valued by individuals as contributing to career development because they facilitate the accumulation of intellectual and social capital – we return to this theme in Chapter 7.

However, the attitudes of both companies and individuals to international mobility are changing, as is discussed below.

The substitutability of mobilities in intra-firm knowledge transfers

There is also a need to think about the substitutability of different forms of mobility, a significant research gap identified by several writers but notably by Salt and Ford (1993: 27) who commented that there are 'some grounds for thinking that migration, secondment, short term assignment and business visits are increasingly substitutable'. Similarly, Hardill (2004: 377) comments that 'cross-border work itself is being transformed, and now includes international commuters, virtual expatriates, frequent flyers and short-term travellers, and so could be considered as being transnational'. There are two important dimensions of this. First, there is substitutability in the types of mobility that transnational companies utilize, whether to manage overseas contracts or effect the circulation of knowledge and corporate practices within the company. As discussed later in this section, Millar and Salt (2008) have identified eight different types of mobility. Other than providing a working typology, their main finding was that firms assembled portfolios of moves, combining different mobility types, in response to both the companies' needs and the changing preferences and aspirations of their workers. These provide different opportunities and constraints for learning and knowledge transfer, so that there are different degrees of substitutability between types of mobility in relation to knowledge transactions.

This leads to the second form of substitutability which can be understood in terms of the concept of 'enfolded mobilities' (Williams 2006b), a concept that emphasizes that mobilities have to be understood as folded into each other over space and time. Individuals engage in different forms of mobilities over time, which impinge on the lives of others, and on their own future mobilities. For example, an initial, relatively long sojourn may establish the knowledge and networks that could make it possible to rely on relatively short trips to that destination in future. Hence professionals may spend a few years in a country, learning much about its nationally-specific knowledge, and developing professional networks. He or she can draw on these in future, either at a distance, or perhaps being refreshed through short return visits. In effect, future learning is made more efficient by previous experiences. To some extent, this extends Wenger's (2000) comments about how occasional spatial proximity is often interwoven with virtual communication in maintaining communities of practice.

In the remainder of this section, we focus specifically on the first type of substitutability, that is between different types of mobility. For example, Perkins (1997: 83–4) recognizes three main forms of international assignment: business trips (less than 31 days' duration per trip), short-term assignments (31 days–12 months), and full assignments (over 12 months). Full assignments conform to traditional definitions of migration, while short-term assignments broadly accord with the notion of temporary migration. A more elaborate and particularly useful eightfold typology is provided by Millar and Salt (2008) which we draw on here: permanent recruitment through the external labour market (ELM) or the internal labour market (ILM); long-term assignments; short-term assignments; commuter assignments; rotators; extended business travel; business travel; and virtual mobility. We elaborate further on these below, while drawing out some of the implications for knowledge transactions.

Permanent recruitment though external or internal labour markets. Permanent recruitment through the external labour market is comparatively rare, but may be pursued if there is a lack of individuals within a company who have sufficient knowledge of a particular country, or who are unwilling to relocate abroad. In some instances, however, there may be individuals who actively seek such relocations – perhaps expatriates who would welcome being relocated to their country of origin, while offering comparative knowledge advantages to the employer (see the discussion of transnationalism in Chapter 4). The latter would depend of course on the length of time they had spent out of the country; that is, on the extent to which their encultured knowledge was up to date. These moves represent 'permanent' migrations.

Long-term assignments. In Millar and Salt's study, most of the companies interviewed considered that long-term assignments were usually of between one and four years in duration. They noted two particular functions of this type of mobility: staff career development and the transfer of knowledge and technological skills. In this instance, the manifest or latent goal is a period of prolonged learning and knowledge accumulation. Some companies may expect staff to acquire international experience (the acquisition of personal knowledge) as being essential for career progression. And they also value the opportunities that these assignments provide for learning about local markets, suppliers, or simply the local business culture.

Short-term assignments. These were usually defined as being of between 3 and 12 months' duration and, therefore, correspond to definitions of temporary migration. For companies, a key distinguishing feature of short-term assignments was the reduced-benefits package that was attached to them, compared to long-term assignments; for example, expenses were not usually paid for the family to relocate for the duration of the assignment. Short-term assignments were also seen as having different functions to long-term ones. Short-term assignments tend to be used for specific tasks including the provision of opportunities for graduate trainees to obtain personal knowledge of other branches of the company, or where an input of their knowledge was required at some stage in the lifecycle of a particular project. This was most likely to be in the decisive early stages, when a company felt that it needed particular expertise to start up and drive forward a venture. They are less effective than long-term assignments in terms of acquiring knowledge of some of the more subtle aspects of local cultures, or in acquiring fluency in the local language.

Commuting. Commuting involves regular mobility between the home base and an international base, usually on a weekly basis, although this could be on a more extended basis in remoter locations. Because of the strains placed on the individuals and their families, these were not usually considered for periods of more than a few months or, at most, for one or two years. While they provide similar opportunities to short-term assignments for the acquisition of encultured and embedded knowledge in the workplace, they were more constrained in relation to the acquisition of knowledge though socializing (due to absences at key socialization events during weekends).

Rotation. Rotators are mostly found in the oil and other extractive industries, where the work is located in particularly challenging and remote locations, such as North Sea oil platforms, or mining in the far reaches of North America. The rotation is often on the basis of working intensively for several weeks, followed by a period of several weeks off duty spent at either the home country, or some other location of choice. Of necessity, this type of mobility provides relatively limited opportunities for acquiring encultured knowledge.

Extended business travel. The companies interviewed differentiated between extended and 'normal' business travel in terms of the expenses that were payable. 30 to 90 days was the most common period over which extended business travel was expected and paid for. This was increasingly a substitute for short-term assignments, partly on the grounds of costs but mostly in order to meet individual preferences. Opportunities for acquiring encultured knowledge were constrained by the shortness of the trip, but also by the probability of living in a highly international milieu such as an international hotel, or a company-owned apartment.

Business travel. Business travel usually involves short trips, which could be made within a day, at one extreme, is usually of a few days' duration, but could extend up to 30 days. This is the most flexible form of travel for most companies, and for individuals. In some ways it is very effective, as there can be steep learning curves at face-to-face meetings that are designed to maximize knowledge transactions. However, they are usually undertaken in an encultured knowledge void, as a result of the classic pattern of flight, taxi, meeting, taxi, flight.

Virtual mobility. This is the reliance on various forms of distanciated communication, ranging from telephone calls and emails to videoconferencing and web cams on PCs. They are least disruptive to the lives of individuals, and cheap to employers, but offer limited opportunities to acquire encultured or embedded knowledge in particular.

The type of mobility that a company elects to use will be driven by a number of factors including the nature of the task, costs, and the willingness of individuals to undertake different types of travel. As noted, the different types of mobility have different implications for knowledge transactions. In turn, these differences reflect several considerations: the time taken to develop positive emotional ties and trust with local workers, the structures for knowledge exchange (e.g. whether formal or informal), and opportunities to acquire knowledge outside the workplace.

In practice, there is likely to be considerable variation in the types of mobility used by companies, depending on both sector and nationality – which in turn is influenced by difference cultural notions of acceptable work patterns and family disruption. For example, both Morley *et al.* (2006: 3) and Scullion (2001) consider than US companies tend to use shorter assignments than UK companies. Erten *et al.* (2006: 40) also comment on changes over time in relation to mobility strategies:

> new forms of international assignments increasingly supplement or partly substitute 'traditional' patterns of expatriation. To be sure, a great number of organizations still use the latter for a number of reasons. . . . However, there is also a growing debate about the significance of expatriation beyond the 'traditional' form: This does not mean the international assignments will diminish in frequency – rather that their form will change. The development of international managers in the future will involve more frequent cross-border job swops, short assignments or assignments to multi-cultural project teams.

These findings are broadly confirmed by Millar and Salt's (2008) study of the aerospace and extractive industries. There is a general shift to shorter-term placements, but more importantly companies are adopting more varied and more sophisticated approaches to managing the mobility of corporate expertise.

There are several reasons for the shifts away from traditional expatriation (see also Box 6.2). Erten *et al.* (2006) emphasize that for companies these include the management of the high costs of mobility. As noted above, there are different levels of benefits and incentives attached to different forms of mobility, and these tend to rise sharply with duration. There are also the costs of failure associated with longer-term placements. For example, McNulty and Tharenou (2006: 18) report that 44 per cent of multinationals in Asia and 63 per cent in Europe report failures in their experiences of expatriation. Moreover, there is also considerable evidence that while most expatriates are relatively happy with the international experiences, they are far more likely to be unhappy with the process of repatriation (Tung 1998). They often find it difficult to settle back into the companies' operations in their country of origin, and there are relatively high levels of resignations, which mean not only a loss of investment, but also knowledge spillovers.

Box 6.2 Democratization of assignments: from expatriates to secondees

Mercer (2006) surveyed some 200 multinational firms across Europe, North America, Latin America, and the Asia & Pacific region to establish the frequency of their global job assignments. They found that the vast majority of multinational companies (84 per cent) utilize short-term assignments. The total numbers of assignments are rising, partly due to the increased level of short-term mobility: 44 per cent of multinational companies increased the number of international assignments from subsidiary to subsidiary between 2003 and 2005.

Firms have moved to reduce their reliance on long-term expatriates. This strategy has become more expensive and also more difficult to implement as many middle-aged employees refuse to sacrifice their family lives in exchange for their careers. Many long-term assignments failed because expatriates and their families were not able to adapt successfully to their new environment. The success of an international assignment also depends on investment in language and cultural training – that is, on acquiring encultured knowledge. While almost all companies recognized this fact, time and cost constraints often lead them to concentrate their efforts on more practical, day-to-day employee support.

International companies have tried to manage problems related to family lives and placement costs by offering alternative forms of

Table 6.1 The nature of short-term assignments

What is the nature of short-term assignments in your firm?	%
Projects	29
Business need	28
Developmental	16
Training	14
Technology initiative	12
Other	1

international mobility. Short-term assignments are generally more cost-effective than long-term ones and they allow companies to transfer knowledge quickly and easily. A common trend is to send individuals on short-term, often 'commuting' assignments lasting about a week. Such assignees or 'secondees' are particularly common in Europe. Improvements in air and rail transport offer higher frequency and lower costs of travel with an increasing density of connected destinations. Many international companies now consider Europe as if it were one country, and have scaled back allowances and benefits to the levels payable for intra-national moves.

The Deloitte Tax Short-term Global Assignment Survey (Deloitte 2006) provides an insight into the driving forces behind international assignments. The survey covered almost 200 companies, and was completed by a mix of Tax Directors and HR personnel across a number of industry groups (see Table 6.1). The Survey revealed a move toward short-term assignment policies that start at less than three months; in contrast, many companies have historically had short-term policies that only commenced at three months.

Source: after Deloitte (2006)

The emergence of more dual careers also means that it is more difficult for couples to uproot and relocate abroad for prolonged stints. Meanwhile, transport and communications changes have made it easier for companies and individuals to opt for shorter-term placements based on rotation or extended business travel, or even commuting. The result (Erten *et al.* 2006: 44) is a shift to 'quasi-expatriation', for example short assignments or international commuting. Whatever the exact form of the mobility, it is clear that both individual workers and companies face considerable challenges if they are to obtain the maximum benefits from such moves.

Migrant knowledge in the organization

Given the importance of knowledge to the organization, a leading priority is to attract and retain knowledgeable workers. This is particularly evident in the high-tech industries where mobility has become common if not routine for many workers. Writing about the IT sector, O'Riain (2004: 222) comments that:

> Negotiating the commitment of highly mobile employees becomes the critical dilemma facing . . . firms . . . the typical career pattern now involves a number of moves between organizations, and there has been a clear shift from internal labor markets to job-hopping between firms.

Employers have a strong incentive to retain (migrant) workers that they have invested in while they acquired encultured and embedded knowledge, while prospective employers are attracted by the prospect of securing not only uncommon knowledge, but also tacit knowledge from their competitors. However, attracting and retaining knowledgeable workers is only the beginning of the challenge for firms. The real key is establishing effective and productive knowledge transactions between workers, and between the new workers and the organizational or collective level.

These challenges are faced in the case of all workers, but are often compounded in the case of international migrants. While migrants seem to have relatively short learning curves for particular competences (Williams and Baláž 2005a), sustained co-learning requires shared norms and understanding, and significant engagement between newcomers and existing personnel. Arguably, in terms of engagement, observation and imitation may be relatively more important with respect to embrained and embodied knowledge, while discourse may be relatively more important for encultured and embedded knowledge.

Co-learning, identities and empathy

Co-learning depends on the willingness of individual workers and organizations to engage with external reference standards (Earl 1990: 742). At the level of the organization and the national state, this is exemplified by the difficulties that workers face in transferring educational credentials between countries. Beyond credentials and nationally-defined skills, the key issue is understanding that knowledge is relational (Bartol and Srivastava 2002). As van der Heijden (2002: 565) argues, 'expertise can only exist by virtue of being respected by knowledgeable people in the organization'. Moreover, 'knowledge gains value when shared with others' (Bertels and Savage 1998: 22). In short, 'sharing' is a disarmingly simple term for a complex process, whereby knowledge transfer and knowledge creation become inseparable from co-learning.

Co-learning flourishes where there are strongly-established norms of trust and cooperation (Empson 2001). Of course, all newcomers to organizations face barriers to co-learning precisely because 'norms, local discourse and other aspects of an organisational or occupational culture are acquired over a significant period of time' (Eraut 2000: 19). However, migrants face particular obstacles to co-learning and knowledge transfer and in the following discussion we draw on Williams (2006a). Observation, imitation and discourse – that is co-learning – are all shaped by the norms of trust and shared understandings. Given that these take time to acquire, it may be particularly difficult for firms or organizations to maximize the knowledge potential of short-term, or temporary, migrants. Overcoming these obstacles – whether in the case of permanent or temporary migrants – is crucial if firms are to realize the 'diversity dividend' from migration.

Diversity is valued by many organizations as a source of learning and knowledge. Amin (2000: 11), for example, argues that 'the infrastructure of soft learning is dissonance and experimentation'. Creative communities actively seek to mobilize difference and counterargument (see also Brown and Duguid 1991), including assigning workers to teams on the basis of nationality differences (Randel 2003). This is an idealized perspective because, in practice, many migrants face significant barriers to knowledge sharing and learning. To some extent this is about positionality, whether in terms of class, gender, ethnicity or migration status. These often intersect and together they are important in determining what people are perceived to know and can do within firms or organizations (Hudson 2004: 450). Moreover, this applies as much to skilled migrants as unskilled migrants (Nagel 2005: 208).

Obstacles to migrant and non-migrant co-learning can also be understood in terms of intercultural communication, the 'symbolic process in which people from different cultures create shared meanings' (Taylor and Osland 2003: 213). Creating shared meanings is difficult where there are strong stereotypes concerning 'the stranger'. Stereotyping – over-generalized expectations and beliefs about the attributes of group membership – increase the likelihood that the voices of strangers will not be heard within an organization. In contrast, cosmopolitanism facilitates inter-cultural exchanges (see also Chapter 5). 'Cosmopolitanism is an orientation, a willingness to engage with the other . . . intellectual and esthetic openness toward divergent cultural experiences, a search for contrasts rather than uniformity' (Hannerz 1996: 104).

Another perspective on inter-cultural communication is provided by Goleman's (1998: 7) notion of 'emotional intelligence', understood as 'managing feelings so that they are expressed appropriately and effectively, enabling people to work together smoothly toward their common goals'. Of the five emotional competences identified by Goleman, empathy is critical because it facilitates 'understanding others' and 'leveraging diversity'. Bogenrieder and Nooteboom (2004), for example, argue that empathy helps in judging trustworthiness. However, 'empathy and identification are generally based on shared experiences, which is one of the obstacles that migrants may face in developing shared identities and empathy with non migrants'.

Co-learning and knowledge transfer are also mediated by social identities, which can be understood as 'the way that identification with a particular social group can be a referent for people to surface certain cognitive assumptions about themselves in relation to others' (Child and Rodrigues 2003: 537). These assumptions – referring back to cosmopolitanism and stereotypes – are critical in how individuals engage – whether positively or negatively – with others. This resonates with Wenger's (1998: 215) comment that learning is 'not just an accumulation of skills and information, but also a process of becoming – to become a certain person or, conversely, to avoid becoming a certain person'. Wenger (2000) also concluded in a later paper that 'a healthy identity' is constituted of 'multimemberships', involving a range of experiences, being open to learning, and identification with broad communities. By extension, therefore, it can be argued that co-learning in workplaces will be facilitated where migrants, and non-migrants, both have 'healthy' identities.

Identities are central to the effectiveness of knowledge transfers and co-learning by international migrants, because nationality and ethnic group membership constitute major social points of reference around which personal identities are worked and reworked (Jenkins 2004: 5). This is increasingly important with the internationalization of labour markets. Nationality and ethnicity are, of course, interwoven with other referents such as gender, age and class, but they are particularly important for international migrants. Companies that seek to maximize co-learning and knowledge transfer understand the importance of social identities, and seek 'to create an affirming work climate for an increasingly multicultural workforce' (Chrobot-Mason and Thomas, 2002: 323–4). Failure to do so may incur a knowledge penalty, by failing to realize the potential of migrant workers.

In part, the importance of social identities depends on the type of knowledge involved. Child and Rodrigues (2003) argue that 'technical knowledge' (about systems and procedures, and strategic understanding) is less likely to be sensitive to social identity, while systemic and strategic knowledge which originate within an organization, are far more identity-sensitive. In any event, there are major challenges for those companies that wish to manage effectively migrant knowledge transactions.

Managing migrant knowledge transactions

If firms are 'repositories of competences, knowledge, and creativity, as sites of invention, innovation and learning' (Amin and Cohendet 2004: 2), how do firms 'harvest' the tacit knowledge of individual migrant workers? Firms face three challenges recruiting individuals with appropriate knowledge; facilitating knowledge exchanges and co-learning among workers; and transferring knowledge to the firm level for redistribution. These pose distinct challenges in relation to migrant workers (see Williams 2006a; 2007a).

Recruitment. All migrants are bearers of personal knowledge, and have learning capacities. However, recruiting migrants poses distinctive challenges for firms in respect of capturing tacit knowledge, particularly their embedded and encultured knowledge. Based on the logic of human capital theories (Chapter 2), migrant workers, and firms, usually incur costs as the newcomers acquire country- or region-specific knowledge. Whether the costs are significant depends on the specific type of knowledge required for particular jobs in particular firms. Writing computer systems software, for example, may require relatively limited encultured knowledge compared to heading up a marketing team. Moreover, in some jobs migrants' distinctive encultured knowledge is highly valued by firms, and is reflected in their recruitment. Examples include tourist guides or receptionists providing services to conational tour groups (Aitken and Hall 2000), and financial services professionals with knowledge of clients and contacts in other countries (McCall 1997). In these cases, human mobility is essential for tacit knowledge transactions, and the challenge for recruitment practices is to recognize those migrants who can transfer or exercise such knowledge effectively (see also Box 6.3).

Box 6.3 Learning-by-hiring

Innovative solutions often originate from borrowing rather than invention. Many firms find it cheaper and faster to source externally available knowledge than to develop competences internally. State-of-the-art knowledge often has a largely tacit nature and is acquired either via experience or learning-by-doing. Human mobility is a way of transferring knowledge, which otherwise would be immovable. When hiring external experts, recipient firms may either value localized and path-dependent strategies (reflecting satisfaction with their current experience) or opt for more innovative, but at the same time more risky, strategies and look for experts with knowledge which is significantly different from their current practices.

Song *et al.* (2003) used US patent and patent-citation data from the global semiconductor industry to study engineers who have moved within the semiconductor industry from US to non-US firms (including moves to both US-based and foreign R&D labs). They also examined their subsequent innovative activities in the hiring firms over a 20-year period (1980–99). The authors identified some 534 patents in 11 patent (technology) classes at the three-digit level that constituted semiconductor-related technology, and then tracked the patenting activities of each of the 180 mobile engineers listed in this sample. Inter-firm mobility was expected to occur if a US-based engineer filed a patent for a US firm, and then subsequently filed a patent for a non-US firm. While patents may be more associated with codified knowledge, flows of codified and tacit knowledge are closely linked and complementary. Patent citations allow exploration of the end points of the knowledge transfer process, regardless of the type of knowledge (codified and/or tacit) involved in its production.

The authors first examined whether the movement of engineers in the sample was associated with significant knowledge transfer from the previous firm to the hiring firm. First patents granted to mobile engineers in the hiring firms were about seven times more likely to cite their previous firms' patents than randomly selected controls from the time period before mobility. This supports the hypothesis of knowledge transfer via learning-by-hiring. Song *et al.* also examined the effect of path dependence on knowledge transfer. Self-citing (occurring when a patent filed by a firm cites another patent from the same firm) was used as a measure of path dependence. When a firm was characterized by stronger path dependence, mobile engineers were less likely to build upon the knowledge in their previous firms. The path-dependence pattern became less important when hired engineers later developed their careers in the hiring firms and were better able to decide their own research direction, incorporating knowledge drawn from their previous firms.

Finally, the authors explored the potential effect of a match between the mobile engineers' ethnicity and the hiring firms' countries of origin. It could be argued that a mobile engineer with the same cultural and linguistic background as the hiring firm may be more likely and able to transfer knowledge. The evidence showed that nationality was important in determining the probability and destination of the migration of experienced engineers. Some 52 out of 64 mobile engineers in the case of Taiwanese, Korean and Japanese firms had the same nationality as their hiring firms. Interestingly, nationality did not seem to substantially affect the level of inter-firm knowledge transfer after mobility had occurred. The significance of diversity therefore remains an open question empirically.

Source: after Song *et al.* (2003)

Some firms may seek to recruit migrants because of their potential value as brokers, or boundary spanners, who can bridge different knowledge communities. This function is not exclusive to migrants and not all migrants have the ability and knowledge to perform this role effectively. However, international migrants potentially can transfer knowledge of different products, processes, markets and social networks. We noted in the previous section how firms value such knowledge and learning in the global financial services sector (Beaverstock 2005). In knowledge communities, companies may welcome mobility because knowledge spillovers resulting from migration and labour force turnover are seen as externalities enjoyed by all firms (see Chapter 5). This may lead to positive and open recruitment practices. But, in some industries, firms seek knowledge monopolies and may focus as much on limiting the mobility of existing workers as on recruiting new (migrant) workers.

Co-learning and knowledge sharing. As argued above, individual knowledge will have little impact on company performance unless shared with other individuals and groups (Nonaka and Takeuchi 1995). Migrant knowledge therefore needs to be harvested by their employers. Companies can design specific learning mechanisms in order to facilitate this (training courses, suggestion boxes etc.) but most tacit knowledge transactions are informal (Ipe 2003: 349), and the key challenge is to create a social environment which is tolerant of diversity and seeks to break down the barriers posed by social identities, positionality and inter-cultural communication differences. In short, informal learning is shaped by firm-level institutions, understood as:

> a common interpretative context based on the visions, values and memories in the form of artifacts, routines and experience which help to ensure that what each employee learns is in some way connected to what the other employees might know or learn.
>
> (Bathelt *et al.* 2004: 34)

It is institutions that define what are sometimes called learning firms or organizations. These try to put in place the opportunities for discussion (project and team meetings etc.) that will facilitate knowledge sharing. But even where a broadly favourable institutional context has been created by a firm, knowledge sharing is ultimately dependent on inter-personal relationships, and intra-group relationships. The fact that the firm is a site of competing interests among individuals and groups (Schoenberger 1997) makes migrant knowledge particularly challenging. However, the most effective firms actively mobilize rather than tolerate such differences (Brown and Duguid 1991) including, we would argue, those arising from migrant status.

Migrants, like most newcomers to firms, begin in peripheral positions within work groups (Lave and Wenger 1991). Effective mobilization involves moving newcomers 'from the domain of stranger toward that of friend' (English-Lucek *et al.* 2002: 97). In part, this involves overcoming relational obstacles around race and nationality or other forms of positionality (Nagel 2005). The 'othering' of migrants, particularly ethnic minorities, generally obstructs knowledge sharing. It can be difficult – and sometimes impossible – for individuals to overcome such obstacles. For example, conscious attempts were made by potential boundary spanners in Japanese banks in London to deepen mutual understanding of each other's knowledge framework. However, these contacts mainly served to emphasize differences, and knowledge transactions remained truncated (Sakai 2000).

Migrants and knowledge management. While firms benefit from informal knowledge transactions between migrants and other workers, they also seek to manage knowledge transfer to the corporate or organizational level. Mobility, including migration, is implicated in two key company strategies for realizing knowledge transfers. First, in epistemic communities (Amin and Cohendet 2004), individuals are brought together to work on particular projects (e.g. new products). These were discussed in the previous section in terms of project ecologies. Where companies have multiple branches, or inter-company collaboration is involved, this may require corporeal mobility, probably alongside virtual communication.

Second – and also discussed in detail in the previous section – multinational companies may, irregularly or routinely, organize various forms of mobility of professional and managerial staff in order to effect knowledge transfers. This depends on the characteristics of the multinational. Bartlett and Ghosal (1989) argued that it is central only to 'transnational' companies that view knowledge as dispersed, and seek to distribute this across the company, irrespective of where it originates. In contrast many multinationals are hierarchical and see the centre as the sources of knowledge, to be channelled down to relatively passive recipients in their international branches. These provide very different roles for international migrants in knowledge transaction. Although we still have relatively limited knowledge of the role of migrants in such companies, we know even less about small and medium-sized firms. Smaller firms, especially those focusing on domestic markets and supply chains, may attach little value to international experiences. Yet in other firms, ethnic diversity is the driving force behind the enterprise as discussed in the final section of this chapter.

Migrants do bring their positionality and their social identities into the workplace, and these mediate knowledge transactions. However, these are not given but, rather, are constantly being made and remade. In other words, migrants are not passive recipients of either positionalities or knowledge-transfer routines (Easterby-Smith and Lyles 2003) that are established by firms. Rather they can contest their positions in knowledge chains. The capacity of an individual, or even a group of, migrants to do so will be contingent on the types of migration: for example, skilled migrants, well-networked in formal and informal professional associations, may be better able to challenge normative mechanisms such as those relating to the recognition of knowledgeable individuals. Box 6.4 presents a case study of international nurses working in a UK hospital.

Box 6.4 International migrant nurses: a case study of a UK hospital

Health care is one of the most widely-studied sectors in relation to the transfer of skills (and knowledge) via migration; for example, see Bach (2003; 2007), Kingma (2006), and Larsen *et al.* (2005). There has been considerable research on health qualifications, not least because these are heavily regulated compared to sectors such as IT or even biotechnology. However, there has been relatively little research on the learning experiences and knowledge transfers effected by migrant nurses. This box reports four in-depth interviews, with three female, and one male, nurses in a major hospital in south-west England. All the interviewees were Filipinos who had been recruited from Singapore where they worked previously. The four had either gone directly to Singapore, or via working in the Middle East, and they specifically saw the former as a stepping-stone to migration to and obtaining a job in the UK.

All four nurses had at least four years experience of working as nurses, but were still required to undergo adaptation training which combined formal courses and on-the-job learning. They found some of the training demanding, and were grateful to be able to refer back to some of their formal training and to knowledge gained from codified forms. Nurse 1 reported that she was often glad that 'I remembered some books I had read, or that I remembered some lectures back in my college days'. The adaptation training was considered to be useful in refreshing or adding to their embrained and embodied knowledge. 'Learning by doing' (Marsick and O'Neill 1999), that is embodied knowledge, was particularly valued.

The desire for practice, as a way of learning, was a recurrent theme in the interviews. Nurse 1, who had worked in theatre in Singapore was frustrated that in the UK hospital she was allocated to the recovery room. Her managers and trainers provided copies of professional articles and papers to read during the long waiting periods, but she considered that such codified knowledge was mostly a waste of time.

Migrant nurses also recognized that they often lacked embedded knowledge, whether of the NHS in general or of the specific hospital that they worked in. To some extent, all mobile workers would face such a challenge when moving between institutions, but this is more marked for international migrants, even when they acknowledge – as in this case – that much of the technical or clinical knowledge employed (whether embrained or codified) is broadly similar to what they had learnt in the Philippines or applied in Saudi Arabia and Singapore. Nurse 2 explained that, for her, the process of learning and adaptation was one of combining codified knowledge with acquired embedded tacit knowledge:

> Knowledge wise, you know how to do pre op, and what to do post op, but paper wise you need to know how to fill out the pre op check list, and what to do after the operation – like one hourly checks at first, and then four hours after the patient is maybe stable.

In addition to formal training courses, all the nurses emphasized the importance of on-the-job training, whereby they acquired all four types of knowledge (Blackler 2002) simply by asking or observing their colleagues. They found most of their colleagues were helpful in this respect, although there were sometimes obstacles in the form of communication. These were occasionally compounded by a reluctance to ask questions, as this would be seen as indicating they were 'not skilled enough to work in the international setting', that is they lacked all the required transferable knowledge.

For some nurses, communication problems could compound the lack of self-confidence, or the ability to understand what was being explained to them. Two main problems were encountered. First, and most common, was the difficulty of understanding particular regional accents. In addition, there were also problems over shared cultural meanings that are inherent in any language – that is, the encultured knowledge that is replete with symbols that are difficult for the 'outsider' to decode. It was most obviously manifest in the use of colloquialisms. This resonates with Blackler *et al.*'s (1998: 75) comment that 'Language does not passively mirror the world, rather speech is a practical act that shapes and negotiates meanings'. This was vividly illustrated by the difficulties caused when the nurses were unable to understand the jokes which formed part of the social glue binding groups of nurses together. The outcome was that international nurses could feel left out and isolated. But it also has implications for learning and for the effectiveness of the nurses because, as Elkjaer (2003: 43) argues, according to social learning theory, language is central to learning, since it is the main way of acting in contemporary organizations.

While the international nurses considered they had all learnt a great deal from their colleagues, they were less certain about the reverse flow of knowledge, that is, whether they had been able to transfer any forms of knowledge to their new colleagues. Nurse 2 explained this in terms of deference, 'I don't feel that I have the right to teach them what I knew . . . I feel I'm just here to learn but not for them to learn from me'. But she then went on to explain that there was also a communication problem, which itself displayed deep cultural differences.

Most believed that they possessed knowledge they could transfer either to their colleagues or to the hospital managers (that is, to the organizational level). Nurse 3 believed that she had a better knowledge of time management (a combination of truncated embedded knowledge, and embrained knowledge) than her British colleagues. Finally, Nurse 1 considered that she had learnt how to better organize operating-theatre trolleys in Singapore than was usual in the UK hospital (again a mixture of embedded, embrained and embodied knowledge), but that 'When it comes to sharing my knowledge, I feel intimidated. I can't explain on that, I'm still developing how to share it'. Yet sharing knowledge is essential if it is to be utilized by organizations.

Source: authors; interviews undertaken by Georgina Adams

Ethnic and migrant enterprises

Ethnic entrepreneurs have been described as 'unsung heroes' (Kloosterman and Rath 2003b: 1) reflecting both their considerable achievements, and the relative lack of research on their activities. The latter has been rectified, to a considerable extent, especially in particular countries such as the USA and the UK (Kloosterman and Rath 2003a; Light and Gold 2000). Here we consider the role of knowledge in such businesses, drawing particularly on notions relating to 'resources' and 'opportunities'.

There are two ways of telling the stories of migrant or ethnic entrepreneurs. First there is the positive or 'good news' story. This argues that by starting their own businesses, ethnic entrepreneurs (not always migrants, it must be noted) can create jobs for themselves and overcome some of the barriers they face as employees – for example, failure to recognize their formal qualifications or, informally, their often diverse knowledge. There may also be outright racism in some contexts, and the othering of 'newcomers' in other settings. In any event, they find it either difficult to enter some industries, or encounter a very low glass ceiling to their endeavours to progress through the hierarchies of particular organizations. One response to these obstacles is to become self-employed, or to establish small businesses that may or may not employ others. These enterprises can be either in the mainstream economy, or in the enclave economy (see also Chapter 2 and the notion of the triple economy). If they succeed they can create jobs for others, whether from their own families or ethnic group, or other ethnic groups. They can also bring a new lease of life to some sectors, such as garment production, which indigenous entrepreneurs have been abandoning. They can make a profit or at least survive in such sectors, because they can draw on specific and distinctive knowledge, social capital and willingness to work long hours.

In contrast, there is the negative or bad news story, which emphasizes that many ethnic entrepreneurs feel compelled to have taken this route by the difficulties they face as employees. They survive in increasingly crowded specialist markets (for example, the multiplication of imitative ethnic restaurants) through long hours, low incomes and intense self-exploitation and exploitation of their family members. Even if they do succeed, they lack the resources, including knowledge, to break out from what turns out to be an economic cul-de-sac in the enclave economy. Reality is far more complex than this simple picture suggests and, for example, migrant and ethnic entrepreneurs are not only found in low-return consumer service sectors. Saxenian (2006), for example, reminds us of the key role that migrant entrepreneurs have played in the development of Silicon Valley. However, Hjerm (2004: 739) found that in Sweden although 'entrepreneurship for immigrants may or may not be positive for the individual . . . it is clear that it is not a successful way to fight economic marginalization and segregation'.

Leaving this debate about how to interpret ethnic or minority entrepreneurship, our main focus here is on the relationship between migration and knowledge. A starting point for understanding this is provided by Waldinger *et al.*'s (1990) so-called 'interactive model'. This situates ethnic enterprise within the complex interactions between opportunity structures and group characteristics. *Opportunities* include market conditions and access to ownership. Market conditions may be generic, for example a demand for convenience stores in low income areas, that indigenous entrepreneurs are not responding to, or ethnic market opportunities. The latter includes an array of services that are demanded by co-ethnics, such as specialist food shops, travel agencies, restaurants and hotels, among others. These implicitly suggest that the possession of very different forms of encultured knowledge by these markets is a key to success: the first implies the acquisition of nationally specific knowledge in the destination, whereas the latter suggests the transfer of knowledge from the country of emigration. Access to ownership is exemplified in two main ways: the availability of vacancies (either as existing owners move out, or vacant property generally becomes available at low costs in run-down inner areas) or through government initiatives to support migrant or ethnic businesses. Specific forms of embedded knowledge are required to access both of these opportunities, that can probably only be acquired over time, or – as discussed below – by drawing on the knowledge of the wider migrant group.

Turning to *group characteristics*, the model identifies predisposing factors and resource mobilization as being important. Predisposing factors include the blocked mobility hypothesis (Bonacich and Modell 1980), which we have already referred to earlier in terms of the obstacles faced as employees. But there are also arguments that migration is highly selective, and that migrants have above-average levels of human capital (or certain forms of knowledge) and aspirations. It is, however, the last of these factors, resource mobilization that is of particular interest here. The resources referred to have close ties to co-ethnics and ethnic networks. These do not apply of course to all migrant groups. However, what they do point to is the resources of the wider group, whether this is capital, social capital, knowledge of how to negotiate the bureaucracy of setting up a business, or of markets and suppliers. Kloosterman and Rath (2003b: 2) explain this:

> Immigrant entrepreneurs may have expert knowledge. . . . In many cases this hard to copy expertise can be based on first-hand knowledge from back home or it can be generated through transnational networks that bridge the country of origin and the sometimes extensive diasporas of a specific group of immigrants.

The resources that are available to ethnic minorities – and to some migrants – can be divided into class and ethnic resources (Light 1984). The class resources are a mixture of the material (for example, property and money) and of knowledge (having access to specific human capital), whereas the ethnic resources emphasize social capital and access to investment from co-ethnics. The work of Portes and Sensenbrenner (1993) is particularly important in this context: they see embeddedness in social networks as grounded in reciprocity transactions, bounded solidarity (facing common adversities), and enforceable trust. These are conditions that facilitate the interchange of resources, whether credit or knowledge. Class resources place more emphasis on individual initiative, whereas ethnic resources stress the collective. Most migrants draw on both types of resources, and Saxenian (1999: 7) comments that in Silicon Valley: 'The region's most successful Chinese and Indian entrepreneurs appear to rely on such ethnic resources while simultaneously integrating into the mainstream technology economy'.

There are, however, limitations to this model, even in advanced capitalist economies, notably an overemphasis on agency and neglect of underlying structural economic changes and institutional frameworks (Kloosterman and Rath 2003b). The possibilities for migrant entrepreneurs are conditioned by national immigration laws and employment regulations, and by structural economic conditions, including the growth of consumerism and the shift to more individualized consumption. Some of these are specific to migrants while others are generic to the economy as a whole. Drawing on Esping-Anderson's (1990: 8) account of the different varieties of capitalist economies in Western Europe, and paraphrasing his arguments, they contend that 'various institutional frameworks also bring about divergent post-industrial employment trajectories by way of path dependent processes'. One reflection of these national differences is the continuing emphasis in much of the European literature on ethnic enterprises in the supposedly relatively low knowledge-intensity consumer services sector, whereas the USA literature increasingly recognizes the role of highly-skilled migrants in high-tech sectors: for example, Min and Bozorgmehr (2003) note the role of highly-skilled Iranians, Indians and Taiwanese, among others, in the professional services in the USA (see Boxes 6.5 and 6.6 for discussion of migrant entrepreneurs in two contrasting sectors). Despite these reservations, Waldinger *et al.*'s (1990) interactive model does provide an useful way to address the role of migrant knowledge in migrant entrepreneurship.

Box 6.5 **Vietnamese entrepreneurs in Slovakia: the value of knowledge in a transition economy**

There were substantial numbers of Vitenamese migrants to Slovakia during the period of state socialism, working mainly in manufacturing or entering as students. The collapse of the state socialist economy after 1989, brought about an economic crisis which was particularly acute for the Vietnamese community. They survived and, to some degree, prospered by moving into market trading. Williams and Baláž (2005b) explored their experiences in terms of Waldinger *et al.*'s (1990) interactive model.

Opportunities. Vietnamese market traders responded to very open *market conditions* in the early transition period in Slovakia, when traditional supply chains were disrupted and real incomes were significantly depressed. They responded by providing low-cost clothing, drawing on their international linkages and knowledge. Their early imports largely originated from Vietnam. However, these goods were not specifically marketed as Vietnamese or Asian, that is, as 'ethnic' products, but simply as low-priced clothing. In terms of *access to ownership*, they benefited from vacancies being available following the collapse of the state socialist economies. In particular, there was very light economic regulation and few limits on 'wild east' retailing as various forms of market trading expanded rapidly in the early 1990s. This is very different to the notion of business vacancy chains in advanced capitalist economies, although some entrepreneurs later took advantage of the sell off of retail units during the privatizations in the mid-1990s. A more important feature of access to ownership was the role of government policies. While positive supportive policies are often critical in advanced capitalist societies, weak and ineffective re-regulation of the emerging market economy facilitated the growth of market trading and petty trading in Slovakia.

Group characteristics. The most important *predisposition factor* was 'blocked mobility' (Bonacich and Modell 1980). The Vietnamese faced high unemployment and poor access to welfare benefits, so they turned to market trading for survival. Their migration was selective but this did not mean they were the more able and enterprising members of their home communities. Before 1989, migration was strictly regulated by bilateral state agreements, and political logic meant that most Vietnamese migrants were from rural North Vietnam. *Resource mobilization* was particularly important, based on close ties among co-ethnics, while their collective identity was fostered by perceived harassment from the police and other authorities, and by segregation, at the level of individual housing blocks, within the main cities. Co-ethnics provided ethnic resources: capital, knowledge about the police and customs, and emotional support. Class-based resources, such as the accumulation of small stocks of human and financial capital under state socialism, were also important: in particular, they developed expertise in making jeans and other clothing items, from textiles imported from Vietnam, working late into the night, even before 1989.

In more recent years the Vietnamese and other market traders have faced continuous, and increasing, competition from other retailing forms, including both international retail chains, and Chinese traders who had even more effective ethnic resources to draw on.

Source: based on Williams and Baláž (2005b)

One particular sub-set of migrant or ethnic entrepreneurs are 'transnational entrepreneurs' (Portes *et al.* 2001: 3), a concept that builds on the notion of transnationalism, as discussed in Chapter 4.

[T]ransnational field is the term coined in the immigration literature to refer to the web of contacts created by immigrants and their home country counterparts who engage in a pattern of repeated back-and-forth movements across national borders in search of economic advantage and political voice.

Box 6.6 **Migrant entrepreneurs in Sweden**

Migrant entrepreneurs in Sweden have relatively high levels of human capital: 46 per cent had some form of higher education compared to 41 per cent of non-migrant entrepreneurs. However, they are over-represented in low-knowledge-intensive consumer services, and under-represented in high-tech firms. One of the reasons for this is that the level of formal qualifications may be less important than the *quality* of human capital or knowledge required for establishing a high-tech firm enterprise. The key qualities in this respect are:

- the importance of prior work experience in the sector;
- the importance of long service in particular high-tech firms, because the required knowledge develops, often unintentionally, through the course of an individual's work history;
- diverse knowledge is often required, and this is classically obtained as engineers and other professionals move from being 'specialists' to 'generalists' during their careers within particular firms;
- such diverse knowledge is easier, and sometimes only possible, to acquire in large firms.

The difficulties for immigrants is that 'ethnic' resources matter far less when establishing high-tech companies than they do in many other types of enterprises:

> [T]he advantages of ethnicity are usually less important than spe-cialized, firm-specific, insider knowledge and contacts. . . . Given its relatively advanced ICT sector and insider networks, work experience and training within Sweden was often far more important for establishing ICT businesses than that gained in home or third countries.

Source: based on Feldman (2006)

The novelty of the concept is the emphasis on the intensity and frequency of contacts across borders, which has been made possible by revolutionary innovations in transportation technology and electronic communications, increasing numbers of immigrants, and the role of governments in both receptor and sending countries in seeking to maximize the benefits from such transnational activities. Under these conditions, transnational activities are becoming increasingly common although the authors caution that as yet they are still the exception rather than the rule. In contrast to the earlier liter-ature on ethnic enterprises, transnational entrepreneurs place relatively greater emphasis on the international as opposed to the national networks of the entrepreneurs. It is precisely their knowledge of two or more places across international borders that gives migrant entrepreneurs a unique resource and competitive advantage. On this basis, Portes *et al.* (2001: 32) contend that in the face of the globalization of capital, and to some extent labour, 'transnational enterprise can give new purchase in the global economy to people of modest means, partially reversing their role as mere sources of wage labor'.

Conclusions: harvesting migrant knowledge

Although the firm or organization is only one of the embedded levels that we need to understand in relation to migrant knowledge transactions, it is the level at which many of the challenges are most clearly articulated and where we start to engage with inter-personal and individual–organizational issues. For individual firms this poses particular challenges in terms of recruitment, creating an environment that facilitates migrant learning and knowledge transfer, and how to transfer such knowledge to the organizational or collective level. Some of these challenges are common to all newcomers to the firm, but others are highly specific to migrants, and relate to inter-cultural communication, positionality and social identities. While these pose significant management challenges, and are conditioned by the wider social construction of migrant positionality and identities, they need to be engaged with if the full potential of migrant knowledge is to be harvested.

The roles and challenges faced by migrants do however vary considerably not only over time and space, but also between company types. International mobility is an important feature of the activities of many multinational companies, and plays a significant role in both harvesting and disseminating knowledge. There is evidence that shorter-term assignments are replacing expatriation and longer sojourns abroad, but it is still very much a case of understanding migration in context of various substitute forms of mobility. Migrant knowledge in ethnic enterprises plays a very different role, constituting one of the key resources that allow such businesses – whether in a positive or negative reading – to create a distinctive niche for migrants.

7 Individual perspectives

The individual: front-stage in knowledge transactions?

We have argued elsewhere for the need to front-stage the individual in ana-lyses of the role of international migration in knowledge transactions (Williams 2006a). This is consistent with the increased focus in economic analysis, in economic geography in particular, on microspaces, drawing attention to people and seeking to avoid the reification of organizations (Ettlinger 2003). It also chimes with Smith's (2001) argument on the need to recognize human agency in the conceptualization of globalization, of what he terms transnationalism 'from below'. In other words, this chapter addresses how flows of knowledge via international migration can be crit-ically shaped by the actions of individuals, as well as shaping their lives and careers.

Care must be taken, however, not to replace the reification of the firm with that of the individual. The individual should not be disembedded from par-ticular social groups. This is particularly important in context of learning and knowledge sharing, as Doornbos *et al.* (2004: 257) remind us: 'Learning results are not only individual but also involve the (tacit) understandings shared by members of a group and inherent in their habitual ways of dealing with situations'. The mobility of individuals also impacts on other members of their households, whether children or elderly parents who may be left behind or uprooted, or spouses who are faced with difficult choices about their own careers. Hardill (2004), for example, writes about the lives of those 'households who live apart' and the tensions within dual-career households. But there is also a need to situate the individual within networks of family and other kin, and friends, who ease and mediate their roles as migrants and in knowledge transactions. In turn, the individual also has to be set in con-text of the particular institutional frameworks of firms, regions and national states. In other words, while arguing for greater recognition of the relatively neglected role of the individual, we do so in context of the multi-layered approach set out in the introduction to this volume. Or, taking a different perspective, as Jenkins (2004: 17) argues, individuals experience the world in terms of three distinct but linked 'orders': the individual order (embodied individuals and what–goes–on–in–their heads), the interaction order (relation-ships amongst people), and the institutional order (structures, organization, and established norms and routines).

There are several reasons why the individual needs to be front-staged in the analysis of the role of migration in knowledge transactions, including the need to avoid excessive generalization, the increasing theoretical stress on human agency, recognition of the dynamism inherent in the life course of migrants, the changing nature of modern economies, greater recognition of the role of individuals in knowledge spillovers, and enhanced understanding of the nature of (knowledgeable) work.

First, there is a paramount need to *avoid undue generalizations*. This is particularly evident in much of the debate about skilled versus unskilled workers, which tends to centre on stereotyped individuals. In reality, of course, there is a continuum of personal knowledge (Polanyi 1975), compounded by the existence of different forms of knowledge (Blackler 2002), that is not easily shoe-horned into any such simplistic dichotomies. As Biao (2005) comments, 'at the end of the day, knowledge exchange must be carried out by individual scientists, and the exchange must become part of the scientist's daily work in order to be effective'. The same applies, of course, to other types of knowledgeable migrants. Greater focus on the individual serves to underline the complexities of knowledge transfers, and helps guard against such stereotypes.

Second, front-staging the individual has to be seen in context of the *increased theoretical emphasis on human agency*. This is consistent with a general shift in migration research from studying aggregates to more qualitative research methods (Massey *et al.* 1993). Beck and Beck-Gernsheim's (2002). Ideas about how relationships are open to interpretation and self-definition in an individualizing world have been particularly influential in migration studies in recent years, notably in studies about student migration (King and Ruiz-Gelices 2003; Baláž and Williams 2004).

Migration and learning studies have also engaged with Giddens' (1884) ideas on structure and human agency (Hardill 2004). The approach is expounded, although using different language, by Goss and Lindquist (1995: 345) in terms of how 'individual interests and actions are not deter-mined by institutions, but individuals draw selectively on institutional rules and resources in pursuit of their interests and inevitably reproduce the social system'. Similarly, social learning theory also emphasizes the inter-relationships between individuals and specific contexts, as Elkjaer (2003: 43) reminds us: 'Individuals are at one and the same time to be regarded as products of their social and cultural history and producing situations mirroring that'. Hodkinson *et al.* (2004) make a similar point in terms of workplace learning, arguing that 'It is not just that each person learns in a context, rather, each person is a reciprocal part of the context, and vice versa'. This is particularly important in this volume, given the emphasis on different forms of knowledge, and on the social situatedness of migrant knowledge transactions and learning.

Third, understanding *the dynamism of the individual life course* is central to understanding the relationship between knowledge transfers and migration. Several commentators have stressed the importance of studying the individual life course in migration studies (for example, King 2002). This is also paralleled by an emphasis on the life course in studies of learning and knowledge transfer, with Billett (1995: 21) commenting that 'the appropriation of knowledge is initially idiosyncratic, being based on the personal histories and epistemologies of individuals'. This leads Hodkinson *et al.* (2004: 8) to advocate a longitudinal or an in-depth ethnographic approach, although they do so in part to guard against 'the danger of exaggerating worker agency'. In terms of this volume, it points to the need to study the way in which particular migration episodes fit into a lifetime of learning and periods of mobility versus non-mobility.

Fourth, *structural and organizational changes in modern economies* are creating more opportunities for individual mobility. Harvey (1982: 380–5) identified the tensions that exist in capitalist economies between the need for aggregate labour mobility to sustain capital accumulation, while individual firms need to try and retain workers (especially if they possess valued knowledge) at particular sites. In more recent years, there has been greater emphasis on increasing labour market flexibility as part of the reworking of the division of labour in the economy. Migrants provide a critical element of such labour market flexibility in some economies. This shift poses challenges for firms that have to compete for labour, and engage newcomers (especially migrants) with the goals and routines of the firm, as noted in Chapter 6.

Fifth, an analysis of the *role of labour mobility in knowledge spillovers* implicitly recognizes the role of individuals. This is a particular challenge in capitalist economies, compared to say state socialist ones, where interworkplace mobility was usually constrained. Smith (2006: 392) argues that labour market researchers have tended to ignore mobility issues because of 'the intellectual and ideological attachment to the power and significance of workplace struggles orchestrated by individual workers, and more importantly workplace unions'. Whatever the reason for this general neglect, it is compounded in relation to studies of international migration which, we have argued throughout this book, plays a key role in knowledge transfers of various types, including knowledge spillovers.

Sixth, recognition of *the nature of work*, including knowledgeable work, reinforces the importance of front-staging the individual. Smith (2006: 389–90) summarizes a key theoretical perspective on this:

> Workers generally know in advance of starting a job the wages they will be paid for their time at work, but not the exact quantity of labour-effort that is required for the particular job. Labour power, what the

employer hires and the worker exchanges, is indeterminate because the precise amount of effort to be extracted cannot be 'fixed' before the engagement of workers, machinery and products for purposeful (profitable within capitalism) action in the labour process. The contract to sell labour power is open-ended, subject to the direction of employers (or supervisory labour) to enforce or create through consent, a definite measure of output from workers over a definite period of time.

This can be rephrased for our purposes by associating effort with knowledge, and the need for firms to secure the consent of individual workers to apply this effectively. Of course, the difficulties and uncertainties involved in negotiating the consent of workers to utilize (and share) knowledge is one reason why firms invest in technology that substitutes codified knowledge in its place. However, as we have argued in Chapter 3, there are limits as to the extent to which tacit knowledge can be translated into codified knowledge, so that a continuing focus on the individual remains central for research in this area.

In summary, the individual level provides an important perspective on knowledge transfer. This is most spectacularly evident in the experiences of exceptionally knowledgeable migrants (see Box 7.1) but also applies to migrants in general as knowledgeable workers.

Box 7.1 Individual migrants: mobility, knowledge and politics

Werner von Braun (1912–77). Von Braun studied at the Technical University of Berlin, where he assisted Herman Oberth, one of the founders of rocketry science and astronautics, in experiments in liquid-fuelled rocket motor tests. In the 1930s and 1940s von Braun developed rocket engines for aircraft and jet-assisted takeoffs, including the long-range A-4 ballistic missile (later renamed the V-2) for the German army. Near the end of the First World War, he led more than 100 of his rocket team members, with over 300 train-car loads of spare V-2 parts, to surrender to the Allied powers. The US Army hired von Braun and his scientists, who were formerly considered war criminals or security threats, to assemble and launch a number of V-2s and develop ballistic missiles. After the launch of Sputnik 1, the USA realized how far it lagged behind the Soviet Union in the emerging Space Race and transferred von Braun to NASA. There he played an instrumental role in the development of the Saturn V rockets, which eventually enabled six teams of astronauts to land on the Moon.

Tsien Hsue-shen (1911–) is one of the founders of Chinese and US space and missile programmes. An immigrant on a Boxer Rebellion Scholarship, Tsien studied at the Massachusetts Institute of Technology and CalTech with Theodor von Karman, a leading American aeronautical engineer. He soon established himself as one of the leading rocket scientists in the US. Towards the end of the Second World War, Tsien was a member of a team of leading scientists who reallocated key documents and personnel from the German aircraft and rocketry programmes to the USA, including von Braun's team. In 1950 Tsien decided to return to China. He was accused of being a Communist party member and detained under house-arrest for five years, while his knowledge on rockets started to become outdated. When released and deported to China in 1955, Tsien became head of the Chinese missile programme. Tsien's team, with help from Russian scientists, started to develop the R-2 rocket, an improved version of the V-2. This research later led to the Dongfeng ballistic missile, successfully launched in 1964. Tsien's team also launched China's first satellite in 1970 and began preparing for human flight in space in the 1980s.

Abdul Qadeer Khan (1935–) is the key founder of Pakistan's nuclear weapons programme. A metallurgy engineer, Khan joined the Physical Dynamics Research Laboratory (FDO) in Amsterdam in 1972. The FDO was a subcontractor for URENCO, the uranium enrichment facility established by the UK, Germany and the Netherlands. About the same time, Pakistan decided to develop a nuclear bomb. Khan allegedly obtained access to classified design documents for centrifuges capable of producing enriched uranium and passed this information to a network of Pakistani intelligence agents. After his return to Pakistan, Khan established the Khan Research Laboratories (KRL) in 1976. The KRL were responsible for designing nuclear bombs and Ghauri ballistic missiles. The first nuclear tests were conducted in 1998. In 2004 Khan confessed on Pakistani national TV that he had provided Iran, Libya and North Korea with designs and technology for mass-destruction weapons.

Source: Boureston (2004); Chang (1996); Ward and Glenn (2005)

Individualization, mobility and lifelong learning

Evans and Rainbird (2002: 9) have argued that 'We need to understand much better the reflexive ways in which people's lives are shaped, bounded or change direction as they engage with education, labour market and workplace organizations'. However, this needs to be seen in the context of increasing labour market flexibility and greater emphasis on individuals taking responsibility for their personal career development, in response to significant changes in the division of labour. As Poell *et al.* (2000: 27) argue: 'The work organization is no longer characterized by a strong Taylorist task division. . . . Employees have also become more and more responsible for their own learning, in order to ensure their employability.'

The need for individuals to take responsibility for their own learning has become something of a mantra in the literature on learning (for example, see Rainbird 2000). This implicates not only formal learning and qualifications but also all the informal experiences of learning across diverse areas of everyday and working lives. As Yang (2003: 108) comments, 'knowledge is learned and accumulated through personal and social life experiences . . . being shaped by both personal inner factors and outside environmental factors'. This chimes with the notion of lifelong learning, expressed by Beckett (2000: 41) in terms of the experience, knowledge and skills that individuals draw on from all areas of their lives, and not only from paid employment. This biographical perspective reinforces the importance of a multi-level perspective on migration that looks beyond the immediate workplace at how knowledge is socially situated, and has changed for individuals over the life course.

Migration is one and, we have argued, an increasingly important means of learning – which can be either a single episode or a series of mobilities over the life course. This has two facets: exposing individuals to different articulations of socially-situated learning, and learning from the experience of migration itself (for example, the knowledge required for successful migration). The scope for migrants to learn in different social situations is, seemingly paradoxically, greater than for non-migrants; while they do not initially possess the encultured and embedded knowledge to fully understand a new setting, their learning is in fact enhanced by their role as 'social outsiders' who do not necessarily fully subscribe to the dominant value system in the destination region or country (Feldman 2006). This makes them more open to a reflexive understanding of a particular setting than insiders who have been deeply embedded within the values of that system, often from birth.

Situated-learning perspectives (Brown and Duguid 1991) are particularly useful in our discussion because they argue that individuals are products of their social and cultural histories whilst also contributing to producing situations that mirror these (Elkjaer 2003). This perspective has two attractions for understanding international migration and learning. First, it places the immediate experiences of the migrants, with respect to knowledge and learning, in context of personal histories of social and spatial mobilities. Second, the emphasis on social and cultural histories directs attention to 'the whole person' because 'experience, knowledge and skills already possessed range over all of a person's life, not just that part of it in paid employment' (Beckett 2000: 41). In other words, workplace learning and knowledge transfer have to be understood in relation to non-workplace experiences. Ettlinger (2003: 152) adopts a similar perspective in researching how multiple rationalities emanate from different spheres of peoples' lives. Paraphrasing such arguments, this paper contends that 'a migrant's knowledge employed in a workplace derives from a kaleidoscope of learning practices that emanate from different spheres of life and different social networks'. The full range of knowledge types is implicated in this argument. For example, Bentley (1998: 104) argues that social skills such as 'spontaneous sociability' (Fukuyama 1995), learnt outside the immediate circle of family and colleagues, are important determinants of employability.

How does the above conceptualization of learning apply specifically to international migrants? While there is a vast literature on migrant communities, and we know a great deal about processes such as social assimilation, there is little research on how practices within migrant communities or families relate to workplace knowledge and learning. However, by way of illustration, we can consider two areas that have been researched: young professionals working in the 'elite meeting grounds' of global cities, and language learning.

Beaverstock (2002; 2005) considers how internationally-mobile workers in the higher-order services constitute transnational elites, who flow into or through global cities, bringing with them well-established cosmopolitan networks, cultural practices and social relations (in our terms, that implies access to encultured and embedded knowledge). The social meeting grounds of global cities are sites such as clubs, restaurants and bars, and these facilitate networking and are critical sites for knowledge transfer/translation and co-learning. They assume different forms in particular cities – bars and clubs in Singapore, lunching in New York, and bars and cafés in London – but have similar functions. Entertaining at home also provides an important arena for knowledge transfer, although the social rules governing this are more complex, as is evident in the differentiation between expatriates and locals in Singapore (Beaverstock 2002: 537).

There is rather more published research on language learning by migrants, another area in which non-workplace learning has significant spillovers (Voydanoff 2001) into the workplace. Not surprisingly, immigrants who live in tightly-bounded ethnic enclaves, with few opportunities to practise the language of the host community, or to venture outside the 'security' of the home and neighbourhood, are likely to have less well-developed foreign-language competence (Chiswick *et al.* 2004; Tomlinson and Egan 2002). Conversely, if there is a favourable environment for language learning at home – e.g. spillovers from children, who have been taught formally at school, to parents – this learning and knowledge can be taken into the workplace. For similar reasons, inter-marriage between immigrants and non-migrants enhances language learning and communication skills (Chiswick and Miller 1995), again with spillovers into the application of knowledge in the workplace. Inter-marriage also facilitates the acquisition of country-specific knowledge (for example, of social customs and institutions) including knowledge of local labour markets (Meng and Gregory 2005). These forms of non-workplace learning have important spillovers in terms of transferring specific encultured and embedded knowledge into the workplace which allows more effective workplace learning and knowledge transactions.

While migration can be understood in context of lifelong learning, there are considerable differences among individuals in their ability to engage effectively with this. For example, writing specifically about entrepreneurs, Coffield (2000: 8) comments that: 'Much could be learned from studying more intensively those learning entrepreneurs, those individuals who make a career of informal learning because learning is part of their wider identities'. Not all migrants fit this model, of course, but it can be argued that migrants – by virtue of the selectivity of migration (by age, education and aspirations), and the challenges posed by migration (a compelling logic of needing to learn) – are particularly likely to engage in lifelong learning (see Box 7.2). Individual migration both facilitates, and is shaped by, lifelong learning.

***Box 7.2* Knowledge migrants to the UK: learning in context**

The UK issued some 60,000 work permits to non-EEA nationals in 2000, mostly to highly-skilled workers. Major increases were reported for work permits in computer services, financial services, and health and medical services. The numbers of permits issued to these three occupational groups increased fivefold, between 1995 and 2000. Against this background, the Department of Trade and Industry and the Home Office commissioned the Institute for Employment Studies (IES) and NOP Business (NOP) to undertake research into the factors that influenced skilled migrant workers to choose the UK as their destination.

The 308 migrants in their sample came from both developed countries (USA, Australia, Canada, New Zealand) and developing ones (India/Pakistan, Far East, Eastern Europe, South Africa). Two factors were common for all migrants: their career ambition and the fact that most were leading a relatively advantaged life in their country of origin. The main migration motives were 'gain experience/knowledge/exposure' (17 per cent), 'career move/career development' (15 per cent), 'better job/career opportunities' (15 per cent) and 'financial reasons' (14 per cent). The rankings of motivations were different for migrants from developed countries and those from developing ones (Figure 7.1). The former were driven mainly by career, culture and travel motives,

Developed countries

(Bar chart, top)

Factor	%
To gain experience/knowledge/exposure	21
Wanted to travel	18
Career move/career development	14
Better opportunities	11
The culture/way of life	10

%

Developed countries

(Bar chart, bottom)

Factor	%
Better pay	18
Better opportunities	17
Career move/career development	16
Studied in UK and decided to stay	15
To gain experience/knowledge/exposure	14

%

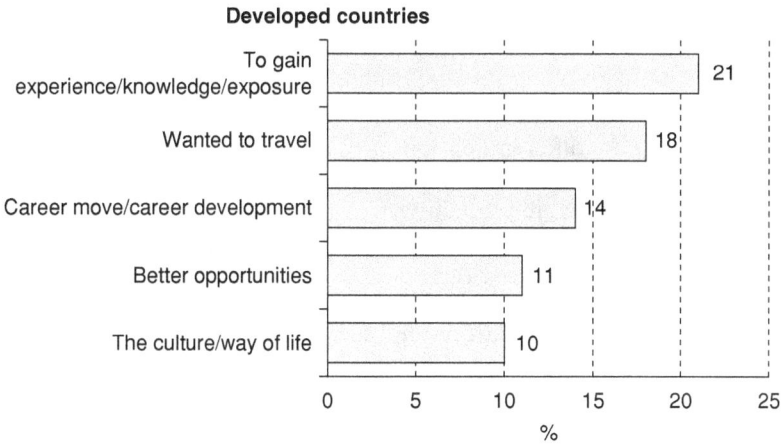

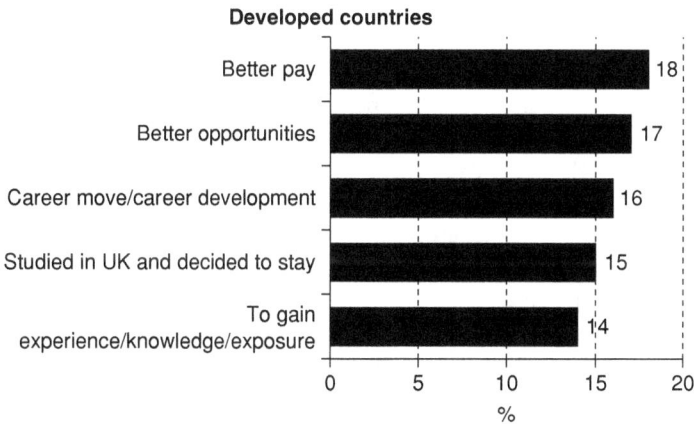

Figure 7.1 Top five attraction factors for skilled migrants to the UK
Source: NOP (2002)

while the latter were motivated more by financial reasons, followed by career motives.

Income differences and earning opportunities were important for some migrants from developing countries, but it was not a dominant factor overall. The skilled migrants wished to develop their careers in global centres of excellence, and/or pursue personal development resulting from travel and experiencing another culture. The Report concludes that the migrants should be thought of as knowledge migrants rather than economic migrants. In other words, migration fitted into a pattern of learning and long-term career advancement, although these were not the only motives, and neither were they homogenous.

Source: after NOP (2002)

Boundaryless careers: upwards and onwards?

International migration has come to play an important role in the careers of many individuals, whether in terms of learning and knowledge acquisition, relocating in different opportunity spaces, or adding international experiences (and knowledge) to their CVs. Careers used to be thought of in terms of being worked through in particular firms and places, and therefore being situated within the same encultured and embedded knowledge contexts. Hardill (2004: 376) expressed this in terms of changing career progression:

> Traditionally the term 'career' implied some long-term progression, a ladder, or linear promotion, within an occupation, or through a series of occupations involving increasing levels of responsibility at each stage. . . . Linear promotion (thereby gaining social mobility) can be achieved through some combination of factors such as length of service, experience, ability and aptitude, and the acquisition of further vocational qualifications . . . as well as utilizing and fostering social capital.

In recent years there has been a shift to more fragmented and discontinuous careers that have involved significant mobility between firms and places, thereby implicating the need to acquire, and emphasizing the value of, different encultured and embedded knowledges. As Barley and Kunda (2004: 22) write, this argument has often been overstated, even in context of high-tech industries, by what they term the 'free agenteer' school of thought which stressed that 'entrepreneurial workers' would regain their independence in the changing world of work. Selling scarce knowledge to eager employers, they had highly mobile careers, and these were the foundations on which they constructed enviable lifestyles:

> To support their claims, free agenteers offered readers stories of contractors who vacationed when and where they chose, who telecommuted to work from exotic places, and who successfully integrated the demands of work and family. Although the heroes and heroines of these tales were usually professional and technical workers, proponents contended that all people could benefit by adopting a similar attitude to work.

Despite such exaggerations, there is no doubt that there has been an increase in careers based on geographical mobility and, as we have argued in this volume, the constitution of such mobility – temporally, spatially and organizationally – mediates knowledge creation and transfer by migrants. One key feature of this is whether migration is constituted as what can be termed *bounded* as opposed to *boundaryless careers* (Williams 2006a).

Bounded careers involve highly-channelled mobility, such as intra-company career or other transfers, or being allocated to work abroad on specific consultancies or contracts. As we noted in Chapters 2 and 4, the pioneering research on skilled labour migration focused on such mobility (Salt 1992), which remains an important feature of the international migration scene, and of corporate organization (Beaverstock and Boardwell 2000; Morgan 2001). The links between company organization and international mobility have been explored by Morgan (2001), who distinguished between two types of international firms. 'Multinationals' are organized along national lines: they have local branches serving local markets, and organize relatively limited human mobility (of any form) among their branches, or between these and the company headquarters. In contrast, 'the global enterprise' is based on transnationally coordinated inter-relationships among different sites. Managers' careers regularly involve (international) mobility between subsidiaries and the HQ. Such mobility facilitates dispersed and multi-directional learning (Morgan 2001: 122).

As noted in Chapter 6, firms value such international placements for several reasons, including the acquisition and transfer of specific knowledge and the development of greater tolerance/openness by their employees through a process of reflective self-learning. Many of these forms of learning contribute to Sennett's (2000) notion of flexible individuals, but also to the role of knowledge brokers, whether as boundary spanners or roamers. They involve not only embrained and embodied knowledge, but also reflexive encultured and embedded knowledge.

Mobile individuals also play an important role in flows of knowledge through extra-firm mobility. Bunnell and Coe (2001) argue that a strong association between individually-centred knowledge and economic rewards has facilitated 'possibly unprecedented mobility' for highly-skilled workers, including international migration. There is anecdotal and fragmented empirical evidence (for example, O'Riain 2004) that large and, at least in some contexts, increasing numbers of workers move as 'free-agent labour migrants': that is, workers who migrate without formal employment contracts, and outside company frameworks. The notion of 'free-agent labour migrants' draws on Opengart and Short's (2002) concept of 'free-agent learners' that, in turn, is based on Kanter's (1995) use of the term 'free agents' (analogous to professional sports terminology). Free-agent labour migrants focus on their long-term employability security within a career model that often stresses geographical mobility.

Free-agent learners move between companies and organizations, seeking lifetime learning, while free-agent international labour migrants also cross borders; they have diverse goals, including those associated with lifelong learning. Free-agent labour migrants are socially diverse, in terms of both their skills and motivations, ranging from young people working abroad as part of a gap year or the 'Big OE' (Big Overseas Experience), to itinerant specialists such as ski instructors, to the plumber or builder who moves from eastern to western Europe in search of employment. From the company perspective, such workers perform diverse roles. Barley and Kunda (2004: 263), for example, distinguish between 'hired guns' (purveyors of just-in-time expertise), and 'gurus' who import skills into the organization that permanent employees lack but need to learn. Such workers (whether or not international) 'epitomized the rhetoric of continual learning'.

International migration, therefore, contributes to what are conceptualized as boundaryless careers (Arthur and Rousseau 1996; Eby 2001). Boundaryless careers are constituted of sequences of jobs across organizations and – in the case of international migrants – across national boundaries. They are facilitated by external networks and contacts, and by a capacity for learning that fosters 'a "free agent" approach to careers, whereby employees are independent from, rather than dependent on, the employing organization' (Eby 2001: 344). Success in the labour market depends on self-knowledge (knowing your strengths), continuously updating knowledge and skills, and renewing social networks (ibid.: 222). Migration is one of the more important signifiers of engagement in the processes that create and sustain employability, as O'Riain (2004) demonstrates in the case of a software-writing team in Ireland. Migration becomes part of networking, creates shared experiences (among mobile workforces), and takes a prominent position in individual CVs. Of course, enhanced worker agency has not eclipsed the importance of structural labour market conditions (Hodkinson *et al.* 2004: 8), but the balance between structure and agency has shifted, leading to more boundaryless careers, and migration is one of the key ways in which this is articulated in labour markets. Box 7.3 provides one example of such a boundaryless career, in the form of an Indian IT specialist.

***Box 7.3* An Indian IT specialist: knowledge, mobility and returns to human capital investment**

Uday, born in 1971, completed a master's degree at Andhara University in 1997. Spurred by the examples of friends and relatives who had already gone abroad, he also determined to make a career in IT working outside India. He went to Singapore, on a tourist visa, to try to find IT employment but found it difficult. On his last day there, he met at a bus stop an Indian who gave him the name of a body-shop company, that is an agency specializing in placing IT workers with other companies on temporary contracts. The company extended his visa and found him three jobs over the next year. When Advanced Technology came to recruit in Singapore in 1999, many of his colleagues signed up to go to North America, but he elected instead for their Australian branch, which he thought would be less competitive. This paid off in that he obtained his permanent residence status in mid-2000, well before any of his colleagues in the USA obtained green cards. Once he had secured his residency, he quit Advanced Technology and moved to a long-term contract with an internet game company.

Source: after Biao (2006: 145)

Employability and social mobility

The key to employability is 'ability to bring a particular kind of knowledge to a task, and be able to collaborate effectively with others to achieve a common task' (Bentley 1998: 103). With structural economic changes, there has been a relative shift to more discontinuous careers. This has meant a shift from employment security to employability security (Opengart and Short 2002: 221) – in other words, the expectation of being able to find new employers, in a labour market where employment is understood to have become increasingly discontinuous. Great emphasis is placed on employability in boundaryless careers – that is on the ability to demonstrate possession of particular knowledge, which makes it possible to change employers with relative ease. This can be seen in either the partly negative sense of being better able to react to greater uncertainty in the labour market, or in the more positive sense of empowering individuals to advance their careers.

International migration can be seen as one feature of how individuals respond to employability issues. They may migrate because there are better opportunities for securing employment in other countries, or because they anticipate that international experience (and knowledge) will make them more employable – whether in destination, origin or other countries. Of course, international migration is undertaken for a variety of reasons, and under varying degrees of duress, and not simply in response to employability challenges. But it can be a source of exceptional learning, as individuals take responsibility for acquiring knowledge and enhancing their employability. In these circumstances, distinctive migrant knowledge (in terms of, for example, encultured knowledge, enhanced reflexivity and having learnt to learn) can become a key component of knowledge accumulation.

However, migration is not automatically a means of enhancing employability. It can also be a stultifying experience, in 'dead-end jobs' with poor learning content. Migrants, because their knowledge is not fully recognized (perhaps because they lack actual, or designed-to-be-exclusionary, national knowledge), or because of obstacles such as labour market regulations, may enter the labour market sub-optimally. Indeed, much of the reasoning behind human capital theories assumes that this will be the case (see Chapter 2). However, it does not automatically follow that the acquisition of nationally-specific human capital will lead to higher wages and occupational mobility for the migrants. The eventual employment outcome depends on whether these initial jobs constitute stepping-stones or labour market entrapments. In the case of 'stepping-stones', there is a gradual matching of knowledge and occupational position, as migrants overcome barriers to using and acquiring knowledge. In the case of 'entrapment', an initial sub-optimal labour market entry has enduring consequences as individuals become 'trapped' in a particular job or labour market segment.

In the context of cycles of migration, as opposed to emigration, the stepping-stones may not be to jobs in the destination, but via return migration to jobs in their country of origin. For example, individual migrants may accept sub-optimal jobs abroad, in order to acquire particular knowledge for which a premium is paid in their countries of origin. This may be embodied or embrained, or (reflexive) encultured or embedded knowledge. As emphasized earlier, the valorization of knowledge is highly place-contingent – it's not only what you know, but where you know it (see Box 7.4).

In the case of stepping-stones, migration may constitute 'significant learning moments' for individuals, whereby they acquire particular forms of knowledge. This is especially so if we look beyond formal qualifications and technical knowledge. Migration can also be a source for acquiring a range of social skills and competences (see Williams and Baláž 2005a). These include the acquisition of self-confidence, networking skills, learning and adaptability competences, and self-reliance. Moreover, as Williams and Baláž (2004) demonstrate for Slovak au pairs in the UK, there is a vast amount of learning and knowledge creation not only in the workplace but also within the private sphere of the employer's home – of, for example, social or language skills. The newly-acquired knowledge may have limited economic value while working as au pairs in the UK, but potentially can be commodified either by working in different jobs in the UK labour market, or as returned migrants in Slovakia. In either case, au pair jobs function as stepping-stones.

Another example of the stepping-stone process is provided by Asian immigrants in Silicon Valley, although the evidence in this well-documented case is somewhat mixed. Without doubt, their higher education – particularly if it was acquired in the USA (that is, nationally-specific and nationally-recognized knowledge) – has often been a stepping-stone into well-paid research jobs in Silicon Valley for many Indian and Chinese migrants (Saxenian 2006). However, they tend to be found disproportionately in professional (technical and engineering) rather than managerial posts. This could be due to technical biases in their education, language difficulties or 'more subtle forms of discrimination or institutional barriers to mobility – the well-known glass ceiling' (Saxenian 2006: 55). If so, this represents a form of labour market entrapment, although in the more privileged reaches of the labour market. In fact there were only relatively minor differences between Asians and Caucasians in terms of their incomes, but significant disparities in relation to their representation in managerial jobs and in terms of upward mobility. Many Asian migrants had responded to these barriers through entrepreneurship (see also Chapter 2): 'Skilled immigrants who experience limits to their professional advancement now have many alternatives including starting their own companies or returning to their home countries' (Saxenian 2006: 81).

Saxenian considers that the return migrants constitute what she terms 'the new Argonauts' (making a comparison with the Argonauts of classical literature), who undertake the risky but rewarding task of starting companies outside the established centres of the high-tech industries. In returning to their countries of origin in order to establish enterprises, or play lead roles as employees in firms in these countries, they had unique advantages as what we have termed migrant knowledge brokers. Box 7.4 provides the example of one such Argonaut.

Box 7.4 **Ronald Chang: a Taiwanese Argonaut**

Ronald Chang was mainland Chinese, brought up in Taiwan, who obtained a PhD in electrical engineering from the University of Southern California in 1975. After several years working for technology companies in the USA, he was recruited to work for Quasel, a Taiwanese semi-conductor producer. He established an R&D centre for them in Silicon Valley and spent the next few years shuttling to and fro across the Pacific. When Quasel failed, he was recruited to run Acer's R&D labs in Taiwan. After five years, he moved to become chief executive of Acer America, a PC manufacturing centre. Six years later, he became chairman of Acer Technology Ventures, a venture capital company that invested in business start-ups in the USA, Taiwan and China.

> Ronald Chang's career is not unusual in Silicon Valley. He has experience in both start-ups and more established technology companies, he had worked as an engineer, manager, executive, and venture investor, and he has been involved with a range of industries from semiconducts and personal computers to peripherals and servers. As he changes jobs, he expands his professional networks and reputation; he also transfers skill and know-how across both firms and business units.
>
> Chang's career is distinctive, however, because it bridges two distant regional economies. He has the language skills and the institutional and cultural know-how to cooperate effectively in both Taiwanese and US business environments, and he maintains relationships with engineers and entrepreneurs in both Silicon Valley and Taiwan.

He is typical of the new Argonauts who build up technological and entrepreneurial capacities in new locations. But they transfer more than just technical knowledge, rather they also transfer what we have termed reflective embedded knowledge, in the form of 'institutional know-how' about creating a positive climate for entrepreneurship.

Source: after Saxenian (2006: 122–3, 326)

Workers in areas such as Silicon Valley tend to have low employment security with individual firms (which tend to have high birth and high death rates) but are highly employable because there are good job opportunities in other firms in the area (Crouch *et al.* 1999), and indeed in emerging centres of production elsewhere in the world. The implications for migration are ambiguous: these conditions may repel or attract individual migrants. Crouch *et al.* also argue that a shift 'from employment to employability' transfers greater responsibility to the individual for acquiring skills and planning career development. This suggests that in practice both bounded and borderless careers, which are explicitly linked to migration, are important in how knowledge communities are constituted and reconstituted in places such as Silicon Valley, and in the transfer of different types of knowledge.

There are also parallels with the acquisition of what Sennett (2000) terms 'flexpertise' – the ability to learn and adapt quickly to changing circumstances. For migrants, this broadly equates with the ability to reflect critically on embodied and encultured knowledge, thereby acquiring distinctive knowledge, or more precisely 'knowledge about knowledge'. It is also about learning how to learn – a challenge that migrants generally encounter in more acute form than non-migrants. The emphasis on flexpertise resonates with the changing nature of career progression that we have already noted in the previous section of this chapter. Gold and Fraser (2002), for example, argue that there has been a shift away from seeing careers as planned linear progressions within organizations, with strong internal labour markets, and long-term employment with individual employers. Instead, careers have changed (Handy 1984) and, increasingly, individuals have 'boundaryless' careers which include 'a range of possible forms that defies traditional employment assumptions' (Arthur and Rousseau 1996: 3). Migration may be interwoven into the flexpertise that allows boundaryless careers to be pursued, while enhancing individual employability – although these relationships are mutually inter-dependent rather than uni-causal.

There are examples of migration either being constituted of stepping-stones, or (in the case of return migration in particular) acting as *a* stepping-stone. However, this potential should not be overstated. The international experiences of many migrants are constituted of deprivation and hardship, with scant opportunities for knowledge acquisition, learning and enhancement of their CVs. Instead, they may become entrapped, with their knowledge being under-valued, and few opportunities to add to this. Entrapment can occur in all three types of labour markets – the primary, secondary and ethnic enclave. Where migrants face obstacles to social or occupational mobility in the primary labour market, this is understood as the 'blocked mobility' thesis (Bonacich and Modell 1980). As we have observed earlier, some migrants circumvent such obstacles through becoming entrepreneurs, or even by a new round of migration or return migration. But if they fail to advance occupationally, by whatever means, then they become entrapped in the labour market of the destination country. And the history of groups such as Puerto Ricans in the USA, Pakistanis in the UK and Moroccans in France, provides broad general evidence of such entrapment – although it may also occur for individuals within groups which generally have experienced upward social or occupational mobility,

Conclusions: migration as individual excursions in learning

The central theme in this chapter is to reinforce the message that knowledge is socially constituted and acquired in specific institutional settings. Whereas we have looked at institutions at the level of the national and regional state, and the firm, in previous chapters, here we have focused specifically on the individual, emphasizing the importance of the family and the community in mediating learning and knowledge transactions. Individuals who move between different institutional settings transfer knowledge, some of which is generic and some of which is place-specific. They also have opportunities to learn both types of knowledge. Of particular interest is embedded and encultured knowledge, which the more reflexive migrants can learn about in a comparative perspective. This allows conclusions to be drawn about how these institutions operate in similar or different way which, as we have argued earlier, is itself a form of commodifiable knowledge. While such learning can be done by moving between institutions within one country, migration between two or more countries, whose institutional settings are *radically* different, offers opportunities not only to acquire and/or transfer uncommon knowledge but also to enhance conceptualization skills.

Migration is also, for individual migrants, a part of their lifelong learning although the extent to which they are conscious of this is highly variable. For some migrants, this is the primary purpose of migration, but for others it is something 'which happens' as a side-effect of, say, the pursuit of higher wages, or being taken as trailing family members by a lead migrant. In turn this learning – and especially 'learning to learn' – can be seen as enhancing their CVs, and their employability. In a world where there are shifts from employment security to employability security, migration can be one of the most potent means to signal that an individual possesses the qualities and motivations required for successful migration, as well as the knowledge and competences acquired from such mobility.

Migration is not however a universal success story. Migrants have very diverse experiences, and their learning can be conditioned by the country, region and firm that they experience, as well as the friends they develop in and out of the workplace, and the neighbourhood they live in. There is a world of difference between the high-flying young migrant who works in the cultural industries, and mixes and learns at work and in his/her neighbourhood from a cosmopolitan and international group, and the older migrant who struggles with the new language, and ends up living in an ethnic enclave, and working in the enclave economy. Between these two idealized types there are of course a variety of positions and experiences. Moreover, migrants are not inevitably condemned to any one particular job or neighbourhood for life. Instead, there can be significant changes over the life course, and migrants can – if unevenly – exert human agency. One of the key questions is whether their initial jobs are stepping-stones or entrapments in the labour market, not only in terms of status and income but also in relation to learning and knowledge. And the commodification of the learning experience – whether of a language, a technical skill, learning ability, communication skills or some other competence – is also dependent on their mobility, and whether they stay, return home or move to another country.

8 Future challenges

International migration and knowledge: key issues

International migration is only one channel for knowledge transfer and learning, and care should be taken not to exaggerate its importance. Indeed, the world labour force is relatively immobile, and most workers never work outside their country of birth. Nevertheless, there has been a fundamental shift in the way that many careers and working lives are played out over space and across international borders, consequent upon changes in transport and communications, in the structure and organization of production and consumption, and uneven economic development. This has been reinforced by the cultural acceptance, and often-positive recognition of mobility as a source of learning, motivated by both economic and non-economic considerations. Moreover, the impact of international migration in knowledge transfer and learning extends beyond the mobile individuals, to non-migrants in areas of origin and destination. Whether through emigration, return, or transnational migrant practices, there are positive and negative spillovers of knowledge and economic impacts for non-migrant workers. Our understanding of the role of international migration in this arena remains partial and fragmented, but a number of key themes have emerged in this volume.

First, tacit knowledge is increasingly recognized as a, and perhaps the most, significant source of competitive advantage. However, tacit knowledge is difficult to transfer and share, and – despite the growth of electronic communication and the growing ease of short-term mobility – international migration remains an important channel of such transfers. The historically high absolute numbers of international migrants in the global economy underlines this.

Second, to some extent international migration is a source of substitute human capital, or a way of replenishing the stock of knowledge in a territorial economy or firm. However, it is also a source of uncommon knowledge. In particular, migrants possess not only embrained and embodied knowledge, but also encultured and embedded knowledge (Blackler 2002), and are potentially knowledge brokers and boundary spanners. Migration can be a source of social diversity that, arguably, is one of the drivers of creativity in modern economies. This is of, course, the positive reading of the story of migration and knowledge. Alternative readings include the negative tale of migrants who fail economically, or who are entrapped in particular labour market segments, or who never acquire sufficient place-specific knowledge in the destination to be able to make use of the full extent of their 'total' knowledge. This is deleterious to their economic prospects while also signalling under-used knowledge resources in a firm or territorial economy.

Third, as indicated above, knowledge is not easily or automatically transferable across borders by individual migrants. Instead, knowledge has to be understood as socially situated – in context of the firm, the community, networks of friends and co-professionals, regions and national spaces. In short, deepening of our understanding of international migration, learning and knowledge, requires linking together our understanding of individuals, social relationships, and institutions in more imaginative ways. Migrants may possess uncommon knowledge, or common knowledge, but the ways in which they utilize these are complex, and place- and organizationally-specific. There are vast differences between not only the young IT high-flier working in a global financial services centre, and the seasonal agricultural migrant worker, but also within these and other groups of so-called skilled and unskilled migrants.

Fourth, while we see migrants as a potential source of knowledge transfer, innovation and enhanced competitiveness, even when the theories of innovation systems, learning regions, knowledge communities, and creativity do acknowledge their role, they do not pay sufficient attention to the constraints faced by individual migrants. These take many forms. They may be due to legal or regulatory restrictions on mobility and the types of jobs available to migrants. Or they may be rooted in stereotyping and racism, or in measures designed to protect the interests of indigenous workers. Migrants are also likely to find it difficult to move from peripheral to core group membership within organizations, and may struggle to assert themselves in the face of sometimes-exclusionary processes of ascription, acceptability and suitability (Jenkins 2004). Social recognition is the first and often the greatest barrier encountered by many migrants. This is not invariably negative, as migrants may sometimes be ascribed to hold more knowledge than they actually do simply on the basis of nationality, as is evident historically for example in relation to colonial administrators or perhaps to French chefs. However, the application of personal knowledge is conditioned by its social recognition by others.

Fifth, knowledge is socially situated and Voydanoff (2001) provides a useful perspective on this, arguing (in a different context) for the need to understand how individuals' seemingly separate lives in the workplace, family and community are interlinked. By extension we can identify the potential for knowledge spillovers between these three spheres. In a sense, enclave communities and workplaces provide truncated learning environments that can be mutually reinforcing for individuals, especially if they are positioned in both and have relatively limited social exchanges and learning opportunities in the wider society and economy. There is research on the integration of individuals in these different spheres. For example, Markova and Black's work (2007) has identified differences among migrants in their interactions with neighbours, other ethnic groups and in the workplace, and some of the key influences on these such as having children, or length of residence in the UK. However, there is still very little research on the linkages between these three spheres, especially in terms of knowledge and learning. One of the main exceptions to this are insights from research, within the human capital perspective, which has indicated how, for example, inter-marriage between migrants and non migrants, or having children at school, enhance the learning experiences of adult migrants, and their individual economic performances.

Sixth, the roles of international migrants have to be understood in context of changing systems of both migration and production. Fielding (1993) provided what was, in many ways, a seminal statement on the link between economic restructuring and international migration, focusing particularly on the shift from Fordism to neo-Fordism. He did not explicitly deal with knowledge issues, but instead identified how a shift to greater labour market flexibility reshaped opportunities and constraints for migrant workers, who also contributed to these structural economic changes. However, the implications of the shift to greater flexibility in the face of intensified competition can be read in apparently contradictory ways. For example, Barley and Kunda (2004) argue that, from an institutionalist perspective, flexibility increases workers' vulnerability and weakens social welfare based on employment; arguably, this also creates a weak workplace environment for learning. In contrast, a 'free labour agents' perspective (Williams 2006a), with its emphasis on human agency, argues that it creates the potential for mobility to be instrumental in the way that individuals can build their own careers. The key to this apparent contradiction lies in the sector-, age-, and place-specificity of workers' experiences. Whereas Barley and Kunda were mostly discussing mobile professionals, in a rather different study Bryant *et al.* (2006) surveyed firms in the food and drink sector in south-west England, and found that the main reasons for employing migrants were as just-in-time workers, or to find staff for difficult-to-fill jobs.

Finally, at the time of writing, international migration has risen to the top of the political agenda in many countries. Research which focuses on knowledge can play an important role in helping to clarify what is often an opaque and distorted understanding of the economic potential of migrants. Hitherto, much of the debate has been structured around the notion of the existence of parallel streams of skilled vs unskilled migrants. In contrast, we have contended that this ignores the existence of learning and knowledge transfer potential and practices among *all migrants*. Firms and territorial economies ignore this at their peril, in context of a competitive and internationalized environment where international migration plays an increasingly important role in determining the winners and losers in a highly-potent but highly-uneven process of knowledge redistribution.

Winners and losers from knowledge transfers via international migration

International migration is often discussed in terms of brain drain vs brain gain, that is, in relation to the net aggregate redistribution of welfare via human mobility. This debate is helpful in highlighting how countries can gain or lose from human capital transfers via international migration flows. However, in reality the picture is more complex than this and there are several factors that mediate the distribution of winners and losers in the process of knowledge transfer and learning.

First, the outcome is of course shaped by the nature of the migration process. In part this is about whether migration is registered or unregistered, which impinges on the opportunities and constraints of employment, and learning for migrants. But this is also a question about whether migration is temporary or permanent. International mobility can be a source of learning which benefits the country of origin, when it is followed by returned migration or repeat circular migration. Similarly, the debate about transnationalism is indicative of increasing potential for some migrants to bridge, and act as knowledge brokers between, different knowledge arenas. Arguably, the greater the economic and cultural differences between places, the greater the potential economic impact of the (uncommon) knowledge transfers between them. However, the greater the gap between places, then the greater the institutional differences, and the more acute the barriers to potential knowledge transfer.

Second, patterns of international migration and knowledge transfer are deeply patterned, indicative of a high degree of path dependency. Migrants are drawn to particular places – whether by higher wages, better learning opportunities, more rewarding working conditions, or the reduced risks attendant on using social networks in these places. At the same time, they contribute to reproducing the economic inequalities that generated these migration flows. They add to the stocks of knowledge in dominant economic places and can play a key role in creating cumulative advantages. However, the processes that are played out are more akin to path-dependent path-creation. This means there is scope for some previously disadvantaged territories to challenge the hegemonic order. Individual free labour market migrants may play this role, as some of the examples in the book indicate. So do national states, as exemplified by the partial success of the Chinese government in encouraging its diaspora to return, or contribute in some other way to the economic performance of the country. However, there are limits to path creation. Knowledge transfer on its own does not lead to innovation and enhanced economic performance. Rather the outcome also depends on the availability of capital, and a favourable institutional framework, as well as stocks of complementary knowledge and a capacity for learning. Therefore, while there is a shifting geography of knowledge and knowledge transfers, this is still characterized by persistent inequalities.

Third, these inequalities are played out at different scales. Migration flows are simultaneously inter-regional and international. In the destination countries, particular regions are usually the main benefactors, with global cities and high-tech regions in developed countries, in particular, being notable winners from such knowledge transfers. Similarly migration from countries of origin may also be highly uneven, as indeed is return, both socially and regionally. Migrants who return, intending to be economically active, are likely to be drawn back to those regions where the economic dynamism offers the greatest returns to their foreign learning, and these may not be the regions where they originated. In other words, transfers of knowledge via international migration and return often contribute to the process of uneven regional developed in different countries (Williams *et al.* 2004). The presence of international migrants and returned migrants does not, of course, guarantee enhanced economic performance but is at least indicative of the potential for uncommon or substitute knowledge transfers that are not available in other regions.

International migration is also associated with winners and losers at the scale of the firm. Ultimately, firms rather than territories are most directly engaged with the economic outcomes of emigration and immigration. Depending on their attitudes to migrant workers, they may perceive these as substitutes – perhaps lower-cost ones – for indigenous workers. They may also stereotype them as fitting particular, often low knowledge-intensity jobs. Or they may see them as bearers of uncommon knowledge, which potentially allows the firm to tap into international rather than just national sources of knowledge and innovation; in other words, a mechanism through which to contest their place in the global. The strategies of firms, together with the organization of the workforce and its cosmopolitanism, will help to determine the distribution of corporate winners and losers. There will also be winners and losers at the individual level. Not all migration results in material and learning benefits, and neither is all international mobility voluntary, however defined. But migration can be a process of learning, that yields knowledge which is commodifiable and career-enhancing for individuals, in context of lifelong learning. The distribution of winners and losers is, however, often strongly polarized, and to some extent unpredictable. While migration can be a stepping-stone for some individuals, for others it can lead to labour market entrapment.

International migration in an uncertain future

In an uncertain economic future, one of the few relative certainties is that international migration is likely to remain of considerable importance. The inherent time-lags involved in demographic shifts mean that most developed economies, and increasing numbers of emerging market economies, are facing relatively predictable ageing profiles. This will make the competition for mobile young workers even more intense in future than at present (Box 8.1), and international migration will constitute one important source of such workers. The logic of uneven capital accumulation will add further to the pressures for increased international migration, although political opposition may also lead to new barriers being raised against large-scale migrations.

Box 8.1 Future migration scenarios

While it is notoriously difficult to estimate future migration flows, some points can be made about the drivers of migration. Putting aside forced migration, then economics, demography and cultural interests are likely to remain major impetuses for population movements. Economics and demographics seem particularly entangled. Current and expected fertility and mortality levels are expected to produce natural population decreases in the developed countries in future (Figure 8.1), and labour shortages in the absence of international migration. The zero-migration scenario predicts a decrease in population numbers of 11.2 per cent in developed countries, and a population increase of 1.8 per cent in developing countries (United Nations 2004b).

How many immigrants will be needed in the developed countries in order to counter these anticipated population decreases? The United

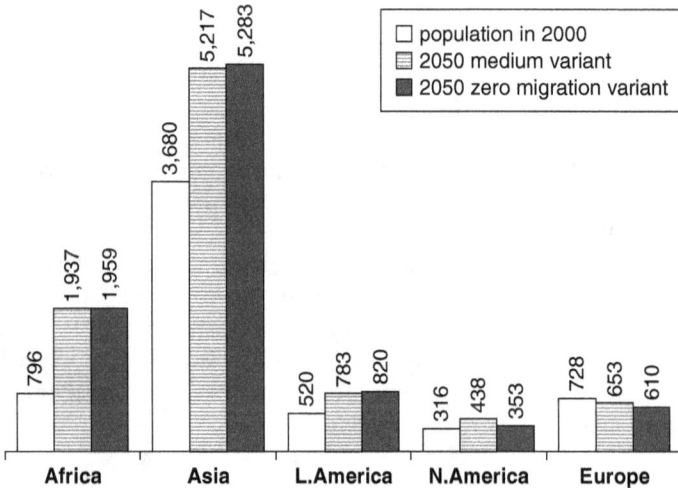

Legend:
- □ population in 2000
- ▤ 2050 medium variant
- ■ 2050 zero migration variant

Africa: 796, 1,937, 1,959
Asia: 3,680, 5,217, 5,283
L.America: 520, 783, 820
N.America: 316, 438, 353
Europe: 728, 653, 610

Figure 8.1 Projected population with and without international migration, 2000 and 2050 (millions)
Source: after United Nations (2002)

Table 8.1 Four scenarios of annual average net immigration for selected countries/regions, 2000–2050 (000s)

Scenario	A	B	C	D
Japan	54	343	647	10,471
Korea	–8	30	129	1,025
USA	1,115	128	359	11,851
EU	680	949	1,588	13,480
UK	136	53	125	1,194
France	75	29	109	1,792
Germany	211	344	487	3,630
Italy	64	251	372	2,268
Russia	50	498	715	5,068

Source: after United Nations (2002)

Nations (2002) World Population Prospects report (see Table 8.1) compares four variants: Scenario A is the medium variant of population development. Scenario B shows the number of migrants required to maintain the size of the existing population. Scenario C shows the number of migrants required to maintain the size of the working population (15–64) constant. Scenario D shows the number of migrants required to keep constant the potential support ratio (i.e., the population aged 15–64 as a proportion of the population aged 65+ years). The required support ratios are very high and probably politically and socially unacceptable for some countries. The EU, for example, would have to receive some 13.5 million immigrants annually during 2000–2050 (some 20 times more than at present) in order to maintain the current level of support ratio. The developed countries therefore face the need to design effective immigration policies, aimed at both skilled and unskilled workers, but will also have to engage with their social integration.

The above analysis, however, only addressed the needs of the more developed countries. Economic, social and cultural changes will also change the demographic profiles, labour needs and propensities to emigrate from and migrate to emerging market economies and, in due course, some of the less developed countries.

Source: after United Nations (2002; 2004b)

The migrations that result from these labour market inequalities will have varying consequences. Some of the migration will fill gaps in, for example, care and cleaning jobs in developed countries, while others will be targeted specifically at migrants because of their high-level tacit knowledge. The latter are the focus of the so-called talent wars (Kuptsch and Fong 2006), and it is likely that countries which wish to succeed in these 'wars' will have to address not only the migrants' aspirations in terms of wages, but also in terms of social integration, learning opportunities, cosmopolitanism and welcoming regulatory regimes.

The resulting mobility will not necessarily involve increased permanent or long-term migration, because there have been relative shifts towards circulation and temporary migration. A certain degree of substitutability will be possible between different forms of mobility, and this has already been significantly enhanced by transport innovations (Box 8.2). There will therefore be a need for more flexible and focused migration policies that address not only the needs of the receiving economies but also of the migrants themselves. In other words, there is a need for a more intelligent approach to migration and related policies which acknowledge the actual and potential contribution of migrants to knowledge creation and transfer.

Box 8.2 **New transport and labour geographies**

Innovations in transport technologies and organization models are contributing to a geographical redefinition of labour markets. The appearance of low-cost airlines (LCAs) and high-speed trains, for example, creates new travel segments, as it offers higher frequency of travel, widens the range of accessible destinations, and lowers transport costs. These innovations enable employers and employees to tap into new labour markets.

Budget air travel has assisted the expansion of a number of smaller firms across Europe. These firms now find it easier and cheaper to meet face-to-face with clients even when they are based in different cities, or countries. The Recruitment Zone, a recruitment consultancy with offices in Edinburgh, Glasgow, Bolton, London and Bratislava in Slovakia, profited from the introduction of EasyJet's services to London, contributing to a doubling of the firm's turnover in 2004 to nearly £15 million, and an increase in staffing from 25 to 60. Prior to the introduction of the EasyJet service, scheduled air fares from Edinburgh to London cost up to £350, compared to the LCA fares of less than £100. The LCA services also enabled the staff to fly between London and its other offices on a regular basis, for short-term work placements or for particular events (Ward 2005).

The cost differentials between LCAs and traditional scheduled airlines are marked. Most travellers choose LCAs for their low fares. A study by the Mercer (2002) consultancy estimated that on short routes (<400 miles), LCAs had half the unit costs of traditional scheduled airlines in the USA. In the UK, operating costs (pence per seat) were 4.19 for EasyJet compared to 9.86 for the British Midlands in 2001.

Low fares, however, are not the only advantage offered by LCAs. For example, a KPMG Survey in Hungary found that 24 per cent of passengers preferred the timetables of low-cost airline to traditional or legacy airlines, while 9 per cent stated that only a low-cost airline offered direct flights for the required destination.

The introduction of LCAs had unexpected consequences for business and labour mobility. Dentists, cosmetic surgeons and spa centres in Poland, Hungary and the Czech and Slovak Republics have benefited from growing numbers of clients from the UK and Germany. The opposite flows have taken Slovak and Polish labour migrants to the UK in particular. Whereas, previously, they returned home twice a year they may now visit their families and friends every month. The outcome is that while international migration is facilitated, so are changes in the forms of international mobility.

Source: *Financial Times* (2005); KPMG (2005); Mercer (2002); Ward (2005); Williams and Baláž (2008a)

Migration policy will need to become more focused on the knowledge that migrants possess, and not just see them as replacement labour. Most of the developed countries are already moving in this direction, with the introduction of various forms of skilled migrant programmes (McLaughlan and Salt 2002), but these are still mostly rooted in conceptions of individual skills rather than of how migrants' knowledge fits into the larger picture of knowledge stocks and transactions, or territorial innovation systems. That is a deficit that can only be addressed by a better theoretical understanding of the relationship between knowledge and migration. It could and should lead to more focused migration policies. However, migration policies only create an enabling framework for knowledge transactions that have to be realized at the inter-personal level within particular organizations. There are therefore major knowledge-management challenges for individual firms if they are to realize such potential. This can be seen in context of the bigger challenge of how to manage social diversity of all types in order to facilitate creativity and learning, but migration poses particular challenges of how to encourage shared learning within organizations.

References

Abella, M. (2006), 'Global competition for skilled workers and consequences', in Kuptsch, C. and Fong, P. E. (eds), *Competing for Global Talent*, Geneva: International Labour Organization, pp. 11–32

Ackers, L. (2004), 'Managing relationships in peripatetic careers: scientific mobility in the European Union', *Women's Studies International Forum*, 27: 189–201

Ackers, L. (2005), 'Moving people and knowledge: scientific mobility in the European Union, *International Migration*, 43: 99–131

Ainley, P. (1993), *Class and Skill: Changing Divisions of Knowledge and Labour*, London: Cassell

Aitken, C. and Hall, C. M. (2000), 'Migrant and foreign skills and their relevance to the tourism industry', *Tourism Geographies: An International Journal of Place, Space and the Environment*, 2: 66–86

Alarcon, R. (1999), 'Recruitment processes among foreign-born engineers and scientists in Silicon Valley', *American Behavioral Scientist*, 42: 1382–97

Almeida, P. and Kogut, B. (1999), 'Localization of knowledge and the mobility of engineers in regional networks', *Management Science*, 45: 905–17

Amin, A. (2000), 'Organisational learning through communities of practice', paper presented at the Millennium Schumpeter Conference, University of Manchester, 28 June–1 July

Amin, A. (2002), 'Spatialities of globalization', *Environment and Planning A*, 34: 385–99

Amin, A. and Cohendet, P. (2004), *Architectures of Knowledge: Firms, Capacities and Communities*, Oxford: Oxford University Press

Amin, A. and Thrift, N. (1994), *Globalization, Institutions and Regional Development in Europe*, Oxford: Oxford University Press

Amin, A. and Thrift, N. (2002), *Cities: Reimagining the Urban*, Cambridge: Polity Press

Andrews, K. M. and Delahaye, B. L. (2000), 'Influences on knowledge processes in organizational learning: the psychological filter', *Journal of Management Studies*, 37: 797–810

Argote, L. and Ingram, P. (2000), 'Knowledge transfer: a basis for competitive advantage in firms', *Organizational Behavior and Human Decision Processes*, 82: 150–69

Arrow, K. J. (1962), 'Economic welfare and the allocation of resources for invention', in Nelson, R. R. (ed.), *The Rate and Direction of Inventive Activity: Economic and Social Factors*, vol. 13, Princeton, NJ: NBER Special Conference Series, Princeton University Press, pp. 609–25

Arthur, M. B. and Rousseau, D. M. (1996), *The Boundaryless Career: A New Employment Principle for a New Organizational Era*, New York: Oxford University Press

Auriol, L. and Sexton, J. (2002), 'Human resources in science and technology: measurement issues and international mobility', in OECD (ed.), *International Migration of the Highly Skilled*, Paris: OECD, pp. 13–38

Bach, S. (2003), *International Migration of Health Workers: Labour and Social Issues*, Geneva: International Labour Organization, Sectoral Activities Programme, Working Paper WP209

Bach, S. (2007), 'Going global? The regulation of nurse migration in the UK', *British Journal of Industrial Relations*, 45: 383–403

Baláž, V. and Williams, A. M. (2004), 'Been there, done that': international student migration and human capital transfers from the UK to Slovakia', *Population, Space and Place*, 10: 217–37

Baláž, V., Williams, A. M. and Kollar, D. (2004), 'Temporary versus permanent youth brain drain: economic implications, *International Migration*, 42: 3–34

Barley, S. and Kunda, G. (2004), *Gurus, Hired Guns, and Warm Bodies: Itinerant Experts in a Knowledge Economy*, Princeton, NJ: Princeton University Press

Barro, R. and Sala-I-Martin, X. (1995), *Economic Growth*, New York: McGraw-Hill

Bartlett, C. A. and Ghosal, S. (1989), *Managing across Borders: The Transnational Solution*, London: Century Business

Bartol, K. M. and Srivastava, A. (2002), 'Encouraging knowledge sharing: the role of organizational reward systems', *Journal of Leadership and Organizational Studies*, 9: 64–77

Bathelt, H., Malmberg, A. and Maskell, P. (2004), 'Clusters and knowledge: local buzz, global pipelines and the process of knowledge creation', *Progress in Human Geography*, (28)1: 31–56

Bauer, T. and Zimmermann, K. F. (1997), 'Looking South and East, labor markets implications of migration in Europe and LDCs', in Memedovic, O., Kuyvenhoven, A. and Molle, W. T. M. (eds), *Globalisation and Labor Markets. Challenges, Adjustment and Policy Responses in the EU and the LDCs*, Dordrecht/Boston/London: Kluwer, pp. 75–103

Beaverstock, J. V. (2002), 'Transnational elites in global cities: British expatriates in Singapore's financial district', *Geoforum*, 33: 525–38

Beaverstock, J. V. (2004), 'Managing across borders: knowledge management and expatriation in professional legal service firms', *Journal of Economic Geography*, 4: 157–79

Beaverstock, J. V. (2005), 'Transnational elites in the city: British highly-skilled inter-company transferees in New York City's financial district', *Journal of Ethnic and Migration Studies*, 31: 245–68

Beaverstock, J. and Boardwell, J. T. (2000), 'Negotiating globalization, transnational corporations and global city financial centres in transient migration studies', *Applied Geography*, 20: 277–304

Beaverstock, J. V., Smith, R. G., Taylor, P. J., Walker, D. R. and Lorimer, H. (2000), 'Globalization and world cities: some measurement methodologies', *Applied Geography*, 20: 43–63

Beck, U. and Beck-Gernsheim, E. (2002), *Individualization*, London: Sage

Becker, G. S. (1964), *Human Capital: A Theoretical and Empirical Analysis, with Special Reference to Education*, Chicago: University of Chicago Press

Beckett, D. (2000), 'Making workplace learning explicit: an epistemology of practice for the whole person', *Westminster Studies in Education*, 23: 41–53

Bentley, T. (1998), *Learning Beyond the Classroom: Education for a Changing World*, London: Routledge

Benton-Short, L., Price, M. D. and Friedman, S. (2005), 'Globalization from below: the ranking of global immigrant cities', *International Journal of Urban and Regional Research*, 29: 945–59

Berger, P. and Luckmann, T. (1966), *The Social Construction of Reality: A Treatise in the Sociology of Knowledge*, London: Penguin

Berry, A. and Soligo, R. (1969), 'Some welfare aspects of international migration', *Journal of Political Economy*, 77: 778–94

Berset, A. and Crevoisier, O. (2006), ' Circulation of competencies and dynamics of regional production systems', *International Journal on Multicultural Societies*, 8(1): 61–83

Bertels, T. and Savage, C. M. (1998), 'Tough questions on knowledge management', in von Krogh, G., Roos, J. and Kleine, D. (eds), *Knowing in Firms: Understanding, Managing, and Measuring Knowledge*, London: Sage Publications, pp. 7–25

Bhagwati, J. and Hamada, K. (1973), 'The brain drain, international integration of markets for professionals and unemployment: a theoretical analysis', *Journal of Development Economics*, 1: 19–42

Biao, X. (2005), *Promoting Knowledge Exchange through Diaspora Networks (The Case of People's Republic of China)*, Oxford: University of Oxford, ESRC Centre on Migration, Policy and Society

Biao, X. (2006), *Global Body Shopping: An Indian Labour System in the Information Technology Age*, Princeton, NJ: Princeton University Press

Billett, S. (1995), 'Workplace learning: its potential and limitations', *Education and Training*, 37: 20–7

Blackler, F. (1995), 'Knowledge, knowledge work, and organizations: an overview and interpretation', *Organization Studies*, 16: 1021–45

Blackler, F. (2002), 'Knowledge, knowledge work and organizations', in Choo, C. W. and Bontis, N. (eds), *The Strategic Management of Intellectual Capital and Organizational Knowledge*, New York: Oxford University Press, pp. 47–62

Blackler, F., Crump, N. and McDonald, S. (1998), 'Knowledge organizations and competition', in Venzin, M., Krogh, G. von and Roos, J. (eds) (1998), *Future Research into Knowledge Management*, Thousand Oaks, CA: Sage, pp. 67–86

Bogenrieder, I. and Nooteboom, B. (2004), 'Learning groups: what types are there? A theoretical analysis and an empirical study in a consultancy firm'. *Organization Studies*, 25: 287–313

Boisot, M. H. (1998), *Knowledge Assets: Securing Competitive Advantage in the Information Economy*, Oxford: Oxford University Press

Bonacich, E. and Modell, J. (1980), *The Economic Basis of Ethnic Solidarity: Small Business in the Japanese American Community*, Berkeley, CA: University of California Press

Borjas, G. J. (1987), 'Self selection and the earnings of immigrants', *American Economic Review*, 77: 531–53

Boureston, J. (2004), 'Tracking the technology', *Nuclear Engineering International*, 31 August

Boyle, P. (2002), 'Population geography: transnational women on the move', *Progress in Human Geography*, 26: 531–43

Boyle, P., Halfacre, K. and Robinson, V. (1998), *Exploring Contemporary Migration.* Harlow: Longman

Bradley, H., Erickson, M., Stephenson, C. and Williams, S. (2000), *Myths at Work*, Oxford: Polity Press

Brettell, C. (2000), 'Theorizing migration in anthropology', in Brettell, C. and Hollifield, J. F. (eds), *Migration Theory: Talking across Disciplines*. London: Routledge

Brown, J. S. and Duguid, P. (1991), 'Organizational learning and communities-of-practice: towards a unified view of working, learning and innovation', *Organizational Science*, 2(1): 40–57

Brown, P. (2001), 'Skill formation in the twenty-first century', in Brown, P., Green, A. and Lauder, H. (2001), *High Skills: Globalisation, Competitiveness and Skill Formation*, Oxford: Oxford University Press, pp. 1–55

Brown, P., Green, A. and Lauder, H. (2001), *High Skills: Globalisation, Competitiveness and Skill Formation*, Oxford: Oxford University Press

Bruin, P. J. M. de (2004), 'Window in the Netherlands. Mapping innovation: regional dimensions of innovation and networking in the Netherlands, *Tijdschrift voor Economische en Sociale Geografie*, 95: 433–40

Bryant, L., Beer, J., Southern, R., Screeton, J. and Inman, A. (2006), *Final Report: A Study of Migrant Workers Employed Within the Food and Drink Sector of the South West of England*, Plymouth: University of Plymouth, Social Research and Regeneration Unit

Bunnell, T. G. and Coe, N. M. (2001), 'Spaces and scales of innovation', *Progress in Human Geography*, 25: 569–89

Burawoy, M. (1979), *Manufacturing Consent*, Chicago: University of Chicago Press

Carrington, W. J. and Detragiache, E. (1998), *How Big Is the Brain Drain?*, Washington, DC: IMF Working Paper, IMF Research Department

Castells, M. (1996), *The Rise of the Network Society (The Information Age: Economy, Society and Culture*, vol. I), Cambridge, MA: Blackwell

Castles, S. and Miller, M. J. (1993), *The Age of Migration*, Basingstoke: Macmillan

Castles, S. and Miller, M. J. (2003), *The Age of Migration: International Population Movements in the Modern World*, third edition, New York: The Guilford Press

Castree, N., Coe, N. M., Ward, K. and Samers, M. (2004), *Spaces of Work: Global Capitalism and the Geographies of Labour*, London: Sage

Cerase, F. P. (1974), 'Expectations and reality: a case study of return migration from the United States to Southern Italy', *International Migration Review*, 8: 245–62

Chang, I. (1996), *Thread of the Silkworm*, New York: Basic Books

Child, J. and Rodrigues, S. (2003), 'Social identity and organizational learning', in Easterby-Smith, M. and Lyles, M. A. (eds), *The Blackwell Handbook of Organizational Learning and Knowledge*, Oxford: Blackwell, 535–55

Chiswick, B. R. (1978), 'The effect of Americanization on the earnings of foreign-born men', *Journal of Political Economy*, 86: 897–921

Chiswick, B. R. and Hatton, T. J. (2003), 'International migration and the integration of labor markets', in Bordo, M., Taylor, A. and Williamson, J. (eds), *Globalization in Historical Perspective*, Cambridge: National Bureau of Economic Research, pp. 65–119

Chiswick, B. R., Lee, Y. L. and Miller, P. W. (2004), *Parents and Children Talk: The Family Dynamics of English Language Proficiency*, London: University College London, Centre for Research and Analysis of Migration, Discussion Paper 3–4

Chiswick, B. R. and Miller, P. W. (1995), 'The endogeneity between language and earnings: international analyses', *Journal of Labor Economics*, 13: 246–88

Chompalov, I., Genuth, J. and Shrum, W. (2002), 'The organization of scientific collaboration', *Research Policy*, 31: 749–67

Chrobot-Mason, D. and Thomas, K. M. (2002), 'Minority employees in majority organizations: the intersection of individual and organizational racial identity in the workplace', *Human Resources Development Review*, 1: 323–44

Clegg, A. (2007), 'In pursuit of the best brains', *Financial Times*, 14 March, p. 12

Cockburn, C. (1983), *Brothers: Male Dominance and Technological Change*, London: Pluto Press

Coffield, F. (2000), 'The structure below the surface: reassessing the significance of informal learning', in Coffield, F. (ed.), *The Necessity of Informal Learning*, Bristol: Polity Press, pp. 1–11

Cohen, R. (1997), *Global Diasporas: An Introduction*, London: Routledge

Concise Oxford Dictionary of Current English (1976), Oxford: Clarendon Press

Cooke, P. and Morgan, K. (1998), *The Associational Economy*, Cambridge: Cambridge University Press

Cronin, V. (2000), *The Wise Man from the West: Matteo Ricci and His Mission to China*, London: Harvill Press (reprint of the 1955 edition)

Crouch, C., Finegold, D. and Sako, M. (1999), *Are Skills the Answer? The Political Economy of Skill Creation in Advanced Industrial Countries*, Oxford: Oxford University Press

Czarniawska, B. (2001), 'Anthropology and organizational learning', in Dierkes, M., Antal, A. B., Child, J. and Nonaka, I. (eds), *Handbook of Organizational Learning and Knowledge*, Oxford: Oxford University Press, pp. 118–35

Damasio, A. (1995), *Descartes' Error: Emotion, Reason, and the Human Brain*, London: Harper Perennial

Dancy, J. (1985), *An Introduction to Contemporary Epistemology*, Oxford: Blackwell

Dassù, M. and Franklin, D. (2005), *Financial Times*, 30 November

Deloitte (2006), *The Deloitte Tax Short-term Global Assignment Survey*, February, Deloitte Development LLC

Docquier, F. and Rapoport, H. (2004), 'Skilled migration: the perspective of developing countries', mimeo, The World Bank

Doornbos, A. J., Bolhuis, S. and Simon, P. R. J. (2004), 'Modeling work-related learning on the basis of intentionality and developmental relatedness: a noneducational perspective', *Human Resource Development Review*, 3: 250–74

Dossani, R. (2002), 'Chinese and Indian engineers and their networks in Silicon Valley', Stanford, CA: Asia/Pacific Research Center

Downes, M. and Thomas, A. S. (1999), 'Managing overseas assignments to build organizational knowledge', *Human Resource Planening*, 22: 33–44

Driouchi, A., Azelmad, M. and Anders, G. C. (2006), 'An econometric analysis of the role of knowledge in economic performance', *Journal of Technology Transfer*, 31: 241–55

Drucker, P. (1993), *Post-Capitalist Society*, London: Butterworth Heinemann

Duany, J. (2002), 'Mobile livelihoods: the sociocultural practices of circular migrants between Puerto Rico and the United States', *International Migration Review*, 36: 355–88

Dustmann, C. (1999), 'Temporary migration, human capital, and language fluency of migrants', *Scandinavian Journal of Economics*, 101(2): 297–314

Dustmann, C. and Fabbri, F. (2002), 'Language proficiency and labour market performance of immigrants in the UK', mimeo, University College London; quoted in Dustmann *et al.* 2003

Dustmann, C., Fabbri, F., Preston, I. and Wadsworth, J. (2003), *Labour Market Performance of Immigrants in the UK Labour Market*, London: Home Office, Online Report 05/03

Dustmann, C. and Weiss, Y. (2007), 'Return migration: theory and empirical evidence from the UK', *British Journal of Industrial Relations*, 45: 236–56

Duvander, A.-Z. E. (2001), 'Do country-specific skills lead to improved labor market positions? An analysis of unemployment and labor market returns to education among immigrants in Sweden', *Work and Occupations*, 28: 210–33

Earl, P. (1990), 'Economics and psychology: a survey', *Economic Journal*, 100: 718–55

Easterby-Smith, M. and Lyles, M. A. (eds) (2003), *The Blackwell Handbook of Organizational Learning and Knowledge*, Oxford: Blackwell

Eby, L. T. (2001), 'The boundaryless career experiences of mobile spouses in dual-earner marriages', *Group & Organization Management*, 26: 343–68

EC (European Commission) (2000), *Push and Pull Factors of International Migration, A Comparative Report*, Luxembourg: Office for Official Publications of the European Communities

Edquist, C. and Johnson, B. (1997), 'Institutions and organizations in systems of innovation', in Edquist, C. (ed.), *Systems of Innovation: Technology, Institutions and Organizations*, London: Pinter, pp. 41–63

Edstrom, A. and Galbraith, J. (1977), 'Transfer of managers as a coordination and control strategy in multinational organizations', *Administrative Science Quarterly*, 22(2): 248–63

Edwards, T. (1998), 'Multinationals, labour management and the process of reverse diffusion: A case study', *International Journal of Human Resource Management*, 9: 696–709

Elkjaer, B. (2003), 'Social learning theory: learning as participation in social processes', in Easterby-Smith, M. and Lyles, M. A. (eds), *The Blackwell Handbook of Organizational Learning and Knowledge*, Oxford: Blackwell, pp. 38–53

Empson, L. (2001), 'Fear of exploitation and fear of contamination: impediments to knowledge transfer in mergers between professional service firms', *Human Relations*, 54: 839–62

English-Lucek, J. A., Darrah, C. N. and Saveri, A. (2002), 'Trusting strangers: work relationships in four high-tech communities', *Information, Communication and Society*, 5: 90–108

Epstein, S. R. (2005), *Transferring Technical Knowledge and Innovating in Europe, c.1200–1800*, Working Papers on 'The Nature of Evidence: How Well Do "Facts" Travel?' No. 01/05, London: London School of Economics, Department of Economic History

Eraut, M. (2000), 'Non-formal learning, implicit learning and tacit knowledge in professional work', in Coffield, F. (ed.), *The Necessity of Informal Learning*, Bristol: Polity Press, pp. 12–31

Erten, C., Schiffinger, M., Mayrhofer, W. and Dunkel, A. (2006), 'The use of foreign assignments and their relationship with economic success and business strategy: a comparative analysis of Northern, Southern and Eastern European countries', in Morley, M. J., Heraty, N. and Collings, D. G. (eds), *New Directions in Expatriate Research*, Basingstoke: Palgrave, pp. 39–63

Esping-Anderson, G. (1990), *The Three Worlds of Welfare Capitalism*, Cambridge: Polity Press

Ettlinger, N. (2003), 'Cultural economic geography and a relational and microspace approach to trusts, rationalities, networks, and change in collaborative workplaces', *Journal of Economic Geography*, 3: 145–71

Etzkowitz, H. and Leydesdorff, L. (2000), 'The dynamics of innovation: from national systems and "mode 2" to a Triple Helix of university-industry-government relations', *Research Policy*, 29: 109–23

EU (2006), *Science and Technology in Europe. Statistical Pocketbook. Data 1990– 2004*, Luxembourg: Office for Official Publications of the European Communities

Evans, K. (2002), 'The challenge of "making learning visible": problems and issues in recognizing tacit skills and key competences', in Evans, K., Hodkinson, P. and Unwin, L. (eds), *Working to Learn: Transforming Learning in the Workplace*, London: Kogan Page, pp. 79–94

Evans, K. and Rainbird, H. (2002), 'The significance of workplace learning for a "learning society"', in Evans, K., Hodkinson, P. and Unwin, L. (eds), *Working to Learn: Transforming Learning in the Workplace*, London: Kogan Page, 7–28

Faist, T. (2000), *The Volume and Dynamics of International Migration and Transnational Social Spaces*, Oxford: Oxford University Press

Faist, T. (2004), 'The border-crossing expansion of social space: concepts, questions and topics', in Faist, T. (ed.), *Transnational Social Spaces: Agents, Networks and Institutions*, Aldershot: Ashgate, pp. 1–34

Feldman, J. M. (2006), 'The limits and possibilities of ethnic entrepreneurship: the case of ICT firms in Sweden', *International Journal on Multicultural Societies*, 8: 84–101

Fermi, L. (1968), *Illustrious Immigrants: The Intellectual Migration from Europe: 1930–41*, Chicago: University of Chicago Press

Fielding, A. (1993), 'Mass migration and economic restructuring', in King, R. (ed.), *Mass Migration in Europe: The Legacy and the Future*, London: Belhaven, pp. 7–18

Financial Times (2005), 'Budget flights give a boost to east Europe's gentle dentists', *Financial Times*, 9 December; available online at: http://search.ft.com/ftArticle? queryText=budget+flights&y=8&aje=true&x=13&id=051209001079&ct=0&nclick_ check=1

Financial Times (2007), 'Call to give chefs star welcome', *Financial Times*, 14 August, p. 2

Findlay, A. M. (2006), 'Brain strain and other social challenges arising from the UK's policy on attracting global talent', in Kuptsch, C. and Fong, P. E. (eds), *Competing for Global Talent*, Geneva: International Labour Organization, pp. 67–86

Findlay, A. M. and Li, F. L. N. (1997), 'An auto-biographical approach to understanding migration: the case of Hong Kong emigrants', *Area*, 29: 34–44

Findlay, A. M. and Stewart, E. (2002), *Skilled Labour Migration from Developing Countries*, Geneva: International Labour Organization, International Migration Programme, International Migration Papers 55

Florida, R. (2002), 'The economic geography of talent', *Annals of the Association of American Geographers*, 92: 743–55

Florida, R. (2005), *Cities and the Creative Class*, New York: Routledge

Flynn, P. M. (1985), *Lowell: A High Technology Success Story*, Waltham, MA: Bentley College

Folbre, N. (2001), *The Invisible Heart: Economics and Family Values*, New York: New Press

Fong, P. E. (2006), 'Foreign talent and development in Singapore', in Kuptsch, C. and Fong, P. E. (eds), *Competing for Global Talent*, Geneva: International Labour Organization, pp. 155–70

Fukuyama, F. (1995), *Trust: The Social Virtues and the Creation of Prosperity*, New York: Free Press

Fukuyama, F. (2001), 'Social capital, civil society and development', *Third World Quarterly – Journal of Emerging Areas*, 22: 7–20

Gaillard, J. and Gaillard, A. M. (1997), 'Introduction: the international mobility of brains. Exodus or circulation?', *Science, Technology and Society*, 2: 195–228

Gamlen, A. (2005), *The Brain Drain Is Dead, Long Live the New Zealand Diaspora*, Oxford: ESRC Centre on Migration, Policy and Society, Working Paper 10

GCIM (Global Commission on International Migration) (2005), *Migration in an Interconnected World: New Directions for Action, Report of the Global Commission on International Migration*, Geneva: GCIM (the report is also available on the Commission's website, www.gcim.org)

Geroski, P. A. (1995), 'Do spillovers undermine the incentive to innovate?', in Dowrick, S. (ed.), *Economic Approaches to Innovation*, Aldershot: Elgar, pp. 76–97

Ghosh, B. (2000), *Return Migration: Journey of Hope and Despair*, Geneva: International Organization for Migration and United Nations

Gibbons, M., Limoges, C., Nowotny, H., Schwartzman, S., Scott, P. and Trow, M. (1994), *The New Production Knowledge: The Dynamics of Science and Research in Contemporary Societies*, London: Sage

Giddens, A. (1884), *The Constitution of Society: Outline of the Theory of Structuration*, Cambridge: Polity Press

Goldfarb, R., Havrylyshyn, O. and Magnum, S. (1984), 'Can remittances compensate for manpower outflows: the case of Philippine physicians', *Journal of Development Economics*, 15: 1–17

Gold, M. and Fraser, J. (2002), 'Managing self-management: successful transitions to portfolio careers', *Work, Employment and Society*, 16: 579–97

Goleman, D. (1998), *Working with Emotional Intelligence*, London: Bloomsbury

Goss, J. and Lindquist, B. (1995), 'Conceptualizing international labour migration: a structuration perspective', *International Migration Review*, 29: 317–51

Gould, W. T. S. (1988), 'Skilled international migrations', *Geoforum*, 19: 381–5

Grabher, G. (2004), 'Learning in projects, remembering in networks? Communality, sociality and connectivity in project ecologies', *European Urban and Regional Studies*, 11: 103–23

Grubel, H. G. and Scott, A. (1977), *The Brain Drain: Determinants, Measurement and Welfare Effects*, Waterloo, ON: Wilfrid Laurier Press

Guarnizo, L. E. (2000), 'Notes on transnationalism', a paper given to the workshop on 'Transnational Migration: Comparative Theory and Research Perspectives', held at Wadham College, University of Oxford. Quoted in Pessar, R. P. and Mahler, S. J., 'Gender and transnational migration', paper given to the conference on 'Transnational Migration: Comparative Perspectives', Princeton University, NJ, 30 June–1 July 2001

Gues, A. P. de (1997), 'The living company', *Harvard Business Review*, 75: 51–9

Handy, C. (1984), *The Future of Work. A Guide to a Changing Society*, Oxford: Basil Blackwell

Hannerz, U. (1996), *Transnational Connections: Culture, People, Places*, London: Routledge

Hardill, I. (2004), 'Transnational living and moving experiences: Intensified mobility and dual-career households', *Population, Space and Place*, 10: 375–89

Hardill, I. and MacDonald, S. (2000), 'Skilled international migration: the experience of nurses in the UK', *Regional Studies*, 34: 681–92

Harvey, D. (1982), *The Limits to Capitalism*, Oxford: Blackwell

Harvey, D. (1990), *The Condition of Postmodernity: An Enquiry into the Origins of Cultural Change*, Cambridge, MA: Blackwell

Hatton, T. J. and Williamson, J. G. (2006), *Global Migration and the World Economy Two Centuries of Policy and Performance*, Cambridge, MA: MIT Press

Heijden, B. I. M. van der (2002), 'Individual career initiatives and their influence upon professional expertise development throughout the career', *International Journal of Training and Development*, 6: 54–79

Henry, N. and Pinch, S. (2000), 'Spatialising knowledge: placing the knowledge community of Motor Sport Valley', *Geoforum*, 31: 191–208

Hess, M. (2004), '"Spatial" relationships? Towards a reconceptualization of embeddedness', *Progress in Human Geography*, 28: 165–86

Hirst, P. and Thompson, G. (1992), 'The problem of "globalization": international economic relations, national economic management and the formation of trading blocs', *Economy and Society*, 21: 357–95

Hjarno, J. (2003), *Illegal Immigrants and Development in Employment in the Labour Markets of the EU*, Aldershot: Ashgate

Hjerm, M. (2004), 'Immigrant entrepreneurship in the Swedish welfare state', *Sociology*, 38: 739–56

Hodkinson, P., Hodkinson, H., Evans, K., Kersh, N., Fuller, A. and Unwin, L. (2004), 'The significance of individual biography in workplace learning', *Studies in the Education of Adults*, 36: 6–24

Hollingsworth, J. R. (2000), 'Theme section: doing institutional analysis: implications for the study of innovations', *Review of International Political Economy*, 7: 595–644

Home Office/DTI (2002), *Knowledge Migrants: the Motivations and Experiences of Professionals in the UK on Work Permits*, London: Home Office/Department of Trade and Industry

Howells, J. (2000), 'Knowledge, innovation and location', in Henry, N., Bryson, J., Daniels, P. and Pollard, J. (eds), *Knowledge, Space, Economy*, London and New York: Routledge

Hudson, R. (2004), 'Conceptualizing economies and their geographies: spaces, flows and circuits', *Progress in Human Geography*, 28: 447–71

Hymer, S. H. (1960), *The International Operations of National Firms: A Study of Direct Foreign Investment*, Cambridge, MA: MIT Press (thesis 1960; published 1976)

IOM (2005), *World Migration 2005: Costs and Benefits of International Migration*, Geneva: International Organization for Migration

Ipe, M. (2003), 'Knowledge sharing in organizations: a conceptual framework', *Human Resource Development Review*, 2: 337–59

Iredale, R. (2001), 'The migration of professionals: theories and typologies', *International Migration*, 39: 7–26

Jacobs, J. (1961), *The Death and Life of Great American Cities*, New York: Random House

Jasso, G. and Rosenzweig, M. R. (1982), 'Estimating the emigration rates of legal immigrants using administrative and survey data: the 1971 cohort of immigrants to the United States', *Demography*, 19: 279–90

Jenkins, R. (2004), *Social Identity*, second edition, London: Routledge

Jessop, R. (2006), 'Why the KBE, why now?', paper presented at the annual conference of the Institute of Advanced Studies, University of Lancaster, September

Jonkers, K. (2004), 'The role of return migrants in the development of Beijing's plant biotechnological cluster', paper prepared for the Second Globelics Conference, 'Innovation Systems and Development: Emerging Opportunities and Challenges', Beijing, China, 16–20 October

Jordan, B. and Düvell, F. (2002), *Irregular Migration: The Dilemmas of Transnational Mobility*, Cheltenham: Edward Elgar

Jordan, B., Stråth, B. and Triandafyllidou, A. (2003), 'Comparing cultures of discretion', *Journal of Ethnic and Migration Studies*, 29: 373–95

Kahneman, D. (2002), 'Maps of bounded rationality: a perspective on intuitive judgment and choice', Prize Lecture, 8 December, in Frangsmyr, T. (ed.), *Lex Prix Nobel*, pp. 416–99. Also accessible online at http://www.nobel.se/economics/laureates/2002/kahnemann-lecture.pdf

Kane, A. A., Argoteb, L. and Levine, J. M. (2005), 'Knowledge transfer between groups via personnel rotation: effects of social identity and knowledge quality', *Organizational Behavior and Human Decision Processes*, 96: 56–71

Kanter, R. M. (1995), 'Nice work if you can get it: the software industry as a model for tomorrow's jobs', *American Prospect*, 23: 52–65

Kapur, D. (2001), 'Diasporas and technology transfer', *Journal of Human Development*, 2: 265–86

Keeble, D., Lawson, C., Smith, H. L., Moore, B. and Wilkinson, F. (1998), *Collective Learning Processes and Inter-firm Networking in Innovative High Technology Regions*, ESRC Centre for Business Research, Working Paper 86

Keep, E., Mayhew, K. and Corney, M. (2002), *Review of the Evidence on the Rate of Return to Employers of Investment in Training and Employer Training Measures*, Warwick and Oxford: Universities of Oxford and Warwick, ESRC Funded Centre on Skills, Knowledge and Organisational Performance, SKOPE Research Paper 34

Khadria, B. (1999), *The Migration of Knowledge Workers: Second-Generation Effects of India's Brain Drain*, New Delhi: Sage Publications

Khadria, B. (2001), 'Shifting paradigms of globalization: the twenty-first century transition towards generics in skilled migration from India', *International Migration*, 39: 45–72

Khaldun, I. (1969), *The Muqaddimah: An Introduction to History*, trans. and intro. Franz Rosenthal, abr. and ed. N. J. Dawood, Princeton, NJ: Bolingen Series, Princeton University Press

Kindleberger, C. P. (1969), *American Business Abroad*, New Haven, CT: Yale University Press

King, R. (1986a), *Return Migration and Regional Economic Problems*, London: Croom Helm

King, R. (1986b), 'Return migration and regional economic development: an overview', in King, R. (ed.), *Return Migration and Regional Economic Problems*, London: Croom Helm, pp. 1–37

King, R. (2002), 'Towards a new map of European migration', *International Journal of Population Geography*, 8: 89–106

King, R. and Ruiz-Gelices, E. (2003), 'International student migration and the European "Year Abroad": effect on European identity and subsequent migration behaviour', *International Journal of Population Geography*, 9: 229–52

Kingma, M. (2006), *Nurses on the Move: Migration and the Global Health Care Economy*, Ithaca, NY: Cornell University Press

Kloosterman, R. and Rath, J. (eds) (2003a), *Immigrant Entrepreneurs: Venturing Abroad in the Age of Globalization*, Oxford: Berg

Kloosterman, R. and Rath, J. (2003b), 'Introduction', in Kloosterman, R. and Rath, J. (eds), *Immigrant Entrepreneurs: Venturing Abroad in the Age of Globalization*, Oxford: Berg

Kofman, E. and Raghuram, P. (2005), 'Gender and skilled migrants: into and beyond the work place', *Geoforum*, 36: 149–54

Kolb, D. A. (1984), *The Experiential Learning: Experience as the Source of Learning and Development*, Upper Saddle River, NJ: Prentice-Hall

Koser, K. and Salt, J. (1997), 'The geography of highly skilled international migration', *International Journal of Population Geography*, 3: 285–303

KPMG (2005), *Low Cost Airlines – What's in it for Tourism?*, Budapest: KMPG

Krieger, H. (2004), *Migration Trends in an Enlarged Europe: Draft*, Dublin: European Foundation for the Improvement of Living and Working Conditions

Kuptsch, C. (2006), 'Students and talent flow – the case of Europe: from castle to harbour', in Kuptsch, C. and Fong, P. E. (eds), *Competing for Global Talent*, Geneva: International Labour Organization, pp. 33–61

Kuptsch, C. and Fong, P. E. (2006), 'Introduction', in Kuptsch, C. and Fong, P. E. (eds), *Competing for Global Talent*, Geneva: International Labour Organization, pp. 1–8

Lapper, R. (2007), 'Globalisation's exiles keep the home fires burning', *Financial Times*, 28 August, p. 7

Larner, J. (1999), *Marco Polo and the Discovery of the World*, New Haven, CT: Yale University Press

Larsen, J. A., Allan, H. T., Bryan, K. and Smith, P. (2005), 'Overseas nurses' motivations for working in the UK: globalization and life politics', *Work, Employment and Society*, 19: 349–68

Lave, J. and Wenger, E. (1991), *Situated Learning: Legitimate Peripheral Participation*, Cambridge: Cambridge University Press

Lee, S. L., Florida, R. and Acs, Z. J. (2004), 'Creativity and entrepreneurship: a regional analysis of new firm formation', *Regional Studies*, 38: 879–92

Li, L., Findlay, A., Jowett, A. and Skeldon, R. (1996), 'Migrating to learn and learning to migrate', *International Journal of Population Geography*, 2: 51–67

Light, D. (1984), 'Immigrant and ethnic enterprise in North America', *Ethnic and Racial Studies*, 7: 195–216

Light, I. and Gold, S. J. (2000), *Ethnic Economies*, San Diego, CA: Academic Press

Lin, N. (2001), *Social Capital*, Cambridge: Cambridge University Press

Lloyd, C. and Payne, J. (2003), 'The political economy of skill and the limits of educational policy', *Journal of Education Policy*, 18: 85–107

Locke, J. (1975), *Essay Concerning Human Understanding*, ed. P. H. Nidditch, Oxford: Clarendon Press

Logan, I. B. (1992), 'The brain drain of professional, technical and kindred workers from developing countries: some lessons from the Africa–USA flow of professionals (1980–1989)', OIM Tenth Seminar on Migration, 'Migration and Development', Geneva, 15–17 September, Report No. 3

Lowell, B. L. and Findlay, A. (2002), *Migration of Highly Skilled Persons from Developing Countries: Impact and Policy Responses*, Geneva: International Labour Organization

Lowell, L., Findlay, A. and Stewart, E. (2004), 'Brain strain', *Asylum and Migration Working Paper 3*, London: Institute of Public Policy Research

Lucas, E. B. (2001), *Diaspora and Development: Highly Skilled Migrants from East Asia*, New York: Report prepared for the World Bank

Lundvall, B. (ed.) (1992), *National Systems of Innovation: Towards a Theory of Innovation and Interactive Learning*, London: Pinter

Lundvall, B. and Johnson, B. (1994), 'The learning economy', *Journal of Industry Studies*, 1: 23–42

McCall, M. (1997), *High Fliers*, Boston, MA: Harvard Business School Press

McCormick, B. and Wahba, J. (2001), 'Overseas work experience, savings and entrepreneurship amongst return migrants to LDCs', *Scottish Journal of Political Economy*, 48: 164–78

McDowell, L., Batnitzky, A. and Dyer, S. (2007), 'Division, segmentation, and interpellation: the embodied labours of migrant workers in a Greater London hotel', *Economic Geography*, 83: 1–25

McGovern, P. (2002), 'Globalization or internationalization? Foreign footballers in the English League, 1946–95', *Sociology*, 36(1): 23–42

McGovern, P. (2007), 'Immigration, labour markets, and employment relations: problems and prospects', *British Journal of Industrial Relations*, 45: 217–35

McLaughlan, G. and Salt, J. (2002), *Migration Policies Toward Highly Skilled Foreign Workers*, London: University College London, Migration Research Unit, Report to the Home Office

McNulty, Y. and Tharenou, P. (2006), 'Moving the research agenda forward on expatriate return on investment', in Morley, M. J., Heraty, N. and Collings, D. G. (eds), *New Directions in Expatriate Research*, Basingstoke: Palgrave, pp. 18–38

Magee, J. and Sugden, J. (2002), 'The world at their feet, professional football and international labour migration', *Journal of Sport & Social Issues*, 26: 421–37

Mahroum, S. (1999), 'Highly skilled globetrotters', in OECD (ed.), *Mobilising Human Resources for Innovation. Proceedings of the OECD Workshop on Science and Technology Labour Markets*, Paris: OECD

Mahroum, S. (2000), 'Highly skilled globetrotters: mapping the international migration of human capital', *R&D Management*, 30: 23–31

Mahroum, S. (2001), 'Europe and the immigration of highly skilled labour', *International Migration*, 39: 27–43

Mahroum, S., Eldridge, C. and Daar, A. (2006), 'Transnational diaspora options: how developing countries could benefit from their emigrant populations', *International Journal on Multicultural Societies*, 8: 25–42

Marcotte, C. and Niosi, J. (2000), 'Technology transfer to China, the issues of knowledge and learning', *Journal of Technology Transfer*, 25: 43–57

Markova, E. and Black, R. (2007), 'East European immigration and community cohesion', Joseph Rowntree Trust website, www.jrf.org.uk

Marsick, V. J. and O'Neil, J. (1999), 'The many faces of action learning', *Management Learning*, 30: 159–76

Martin, P. L. (2006), 'Competing for global talent: the US experience', in Kuptsch, C. and Fong, P. E. (eds), *Competing for Global Talent*, Geneva: International Labour Organization, pp. 87–105

Martin, P. and Midgley, E. (2003), 'Immigration: shaping and reshaping America', *Population Bulletin*, June

Maskell, P. and Malmberg, A. (1999), 'Localised learning and industrial competitiveness', *Cambridge Journal of Economics*, 23: 167–85

Massey, D. (1994), *Place, Space and Gender*, Minneapolis, MN: University of Minnesota Press

Massey, D. S., Arango, J., Hugo, G., Kouaouci, A., Pellegrino, A. and Taylor, J. E. (1993), 'Theories of international migration: a review and appraisal', *Population and Development Review*, 19: 431–67

Matthews, G. and Ruhs, M. (2007), *Are You Being Served? Employer Demand for Migrant Labour in the UK's Hospitality Sector*, Oxford: University of Oxford, COMPAS Working Paper 07-51

Meng, X. and Gregory, R. G. (2005), 'Intermarriage and the economic assimilation of immigrants', *Journal of Labor Economics*, 23, 135–75

Mercer (2002), *Impact of Low Cost Airlines*, New York: Mercer Management Consulting

Mercer (2006), *International Assignments Survey 2005/2006*, New York: Mercer Management Consulting

Meyer, J.-B. (2001), 'Network approach versus brain drain: lessons from the diaspora', *International Migration*, 39: 91–110

Meyer, J.-B., Bernal, D., Charum, J., Gaillard, J., Granes, J., Leon, J., Montenegro, A., Morales, A., Murcia, C., Narvaez Berthelemot, N., Parrado, L. and Schlemmer, B. (1997), 'Turning brain drain into brain gain: the Colombian experience of the diaspora option', *Science, Technology and Society*, 2: 285–315

Meyer, J.-B. and Brown, M. (1999), 'A new approach to the brain drain', Discussion paper, No. 41, World Conference on Science, UNESCO-ICSU, Budapest, June–July

Meyer, J.-B., Kaplan, D. and Caran, J. (2001), *Scientific Nomadism and the New Geopolitics of Knowledge*, Oxford: UNESCO

Meyer, J.-B. and Wattiaux, J.-P. (2006), 'Diaspora knowledge networks: vanishing doubts and increasing evidence, *International Journal on Multicultural Societies*, 8: 4–24

Millar, J. and Salt, J. (2007), 'In whose interests? Migration in an interconnected world economy', *Population, Space and Place*, 13: 41–58

Millar, J. and Salt, J. (2008), 'Portfolios of mobility: the movement of expertise in transnational corporations in two sectors – aerospace and extractive industries', *Global Networks*, 8(1): 25–50

Min, P. G. and Bozorgmehr, M. (2003), 'The United States: the entrepreneurial cutting edge', in Kloosterman, R. and Rath, J. (eds), *Immigrant Entrepreneurs: Venturing Abroad in the Age of Globalization*, Oxford: Berg, pp. 17–37

Møen, J. (2005), 'Is mobility of technical personnel a source of R&D spillovers?', *Journal of Labor Economics*, 23: 81–114

Morgan, G. (2001), 'Transnational communities and business systems', *Global Networks*, 1: 113–30

Morley, M. J., Heraty, N. and Collings, D. G. (2006), 'Introduction: new directions in expatriate research', in Morley, M. J., Heraty, N. and Collings, D. G. (eds), *New Directions in Expatriate Research*, Basingstoke: Palgrave, pp. 1–17

Mountford, A. (1994), *Can a Brain Drain Be Good for Growth?*, Tilburg, Netherlands: Tilburg University, Center for Economic Research

Munir, K. (2002), 'Being different: how normative and cognitive aspects of institutional environments influence technology transfer', *Human Relations*, 55(12): 1403–28

Nadler, J., Thompson, L. and Boven, L. van (2003), 'Learning negotiation skills: four models of knowledge creation and transfer', *Management Science*, 49: 529–40

Nagel, C. (2005), 'Skilled migration in global cities from "Other" perspectives: British Arabs, identity politics, and local embededdness', *Geoforum*, 36: 197–210

Nählinder, J. and Hommen, L. (2002), *Employment and Innovation in Services: Knowledge Intensive Business Services in Sweden*, London: Birkbeck College, University of London, Clore Management Centre

NBSC (National Bureau of Statistics of China) (2006), *China Statistical Yearbook 2005*, Beijing: China Statistics Press, Beijing Info Press, Chs 21–8

Neatrour, S. and Williams, J. (2002), *The New Football Economics* (Sir Norman Chester Centre for Football Research, Fact Sheet 10), Leicester: University of Leicester

Nelson, R. (1993), *National Innovation Systems: A Comparative Analysis*, Oxford: Oxford University Press

Newbold, K. B. (2001), 'Counting migrants and migrations: comparing lifetime and fixed-interval return and onward migration', *Economic Geography*, 77: 23–40

Nonaka, I. and Takeuchi, H. (1995), *The Knowledge Creating Company: How the Japanese Companies Create the Dynamics of Innovation*, New York: Oxford University Press

Nonaka, I., Toyama, R. and Nagata, A. (2000), 'A firm as a knowledge-creating entity: a new perspective on the theory of the firm', *Industrial and Corporate Change*, 9: 1–20

Nonaka, I., Toyama, R. and Byosière, P. (2001), 'A theory of organizational knowledge creation: understanding the dynamic process of creating knowledge', in Dierkes, M., Antal, A. B., Child, J. and Nonaka, I. (eds), *Handbook of Organizational Learning and Knowledge*, Oxford: Oxford University Press, pp. 491–517

Noon, M. and Blyton, P. (1997), *The Realities of Work*, Basingstoke: Macmillan

NOP Business/Institute for Employment Studies (2002), *Knowledge Migrants: The Motivations and Experiences of Professionals in the UK on Work Permits: Final Report*, London: Home Office and Department of Trade and Industry

OECD (1996), 'The knowledge-based economy', in OECD (ed.), *STI Outlook*, Paris: OECD

OECD (2002), *International Mobility of the Highly Skilled*, Paris: OECD

OECD (2004), *Education at a Glance 2004 – Tables*: 'Indicator C3: Foreign Students in Tertiary Education': 'Table C3.2. Foreign students in tertiary education by country of origin (2002)', Paris: OECD; available online at: http://www.oecd.org/document/11/0,3343,en_2649_39263238_33712011_1_1_1_1,00.html

OECD (2006), *The OECD Science, Technology and Industry Scoreboard*, Paris: OECD

Oinas, P. (2000), 'Distance and learning: does proximity matter?', in Boekma, F., Bakkers, S. and Rutten, R. (eds), *Knowledge, Innovation and Economic Growth: The Theory and Practice of Learning Regions*, Cheltenham: Edward Elgar, pp. 57–69

Olesen, H. (2002), 'Migration, return and development', *International Migration*, 40: 125–50

Opengart, R. and Short, D. C. (2002), 'Free agent learners: the new career model and its impact on human resource development', *International Journal of Lifelong Education*, 21: 220–33

O'Riain, S. (2004), 'Net-working for a living: Irish software developers in the global workplace', in Amin, A. and Thrift, N. (eds), *The Blackwell Cultural Economy Reader*, Oxford: Blackwell, pp. 15–39

Ossman, S. (2004), 'Studies in serial migration', *International Migration*, 42: 111–21

Park, S. O. (2004), 'Knowledge, networks and regional development in the periphery in the internet era', *Progress in Human Geography*, 28: 283–86

Payne, J. (2000), 'The unbearable lightness of skill: the changing meaning of skill in UK policy discourses and some implications for education and training', *Journal of Education Policy*, 15: 353–69

Pellerin, H. (1999), 'The cart before the horse? The coordination of migration policies in the Americas and the neoliberal economic project of integration', *Review of International Political Economy*, 6: 468–93

Perkins, S. J. (1997), *Internationalization: the People Business*, London: Kogan Page

Piore, M. (1979), *Birds of Passage: Migrant Labor in Industrial Societies*, Cambridge: Cambridge University Press

Poell, R. F., Chivers, G. E., Krogt, F. J. van der, and Wildemeersch, D. A. (2000), 'Learning-network theory: organizing the dynamic relationships between learning and work', *Management Learning*, 31: 25–49

Polanyi, M. (1958), *Personal Knowledge*, London: Routledge and Kegan Paul

Polanyi, M. (1966), *The Tacit Dimension*, London: Routledge and Kegan Paul

Polanyi, M. (1975), 'Personal knowledge', in Polanyi, M. and Prosch, H. (eds), *Meaning*, Chicago: University of Chicago Press, pp. 22–45

Porter, M. E. (2003), 'The economic performance of regions', *Regional Studies*, 37: 549–78

Portes, A. (1997), 'Immigration theory for a new century: some problems and opportunities', *International Migration Review*, 31: 799–827

Portes, A. and Bach, R. L. (1985), *Latin Journey: Cuban and Mexican Immigrants in the United States*, Berkeley and Los Angeles: University of California Press

Portes, A., Guarnizo, L. E. and Landolt, P. (1999), 'The study of transnationalism: pitfalls and promise of an emergent research field', *Ethnic and Racial Studies*, 22: 219–37

Portes, A., Haller, W. and Guarnizo, L. E. (2001), *Transnational Entrepreneurs: The Emergence and Determinants of an Alternative Form of Immigrant Economic Adaptation*, Oxford: University of Oxford, Transnational Communities Programame, Working Paper 01-05

Portes, A. and Sensenbrenner, J. (1993), 'Embeddedness and immigration: notes on the social determinants of economic action', *American Journal of Sociology*, 98: 1320–50

Rabi, M. M. (1967), *The Political Theory of Ibn Khaldun*, Leiden: E. J. Brill

Raghuram, P. (2002), 'Negotiating Skilled Migration', ESRC Seminar: 'Beyond Contact: Borders, Bodies and Bonds' (Warwick University), February

Raghuram, P. (2004), 'The difference that skills make: gender, family migration strategies and regulated labour markets', *Journal of Ethnic and Migration Studies*, 30: 303–21

Rainbird, H. (2000), 'Skilling the unskilled: access to work-based learning and the lifelong learning agenda', *Journal of Education and Work*, 13: 183–97

Randel, A. E. (2003), 'The salience of culture in multinational teams and its relation to team citizenship behavior', *International Journal of Cross Cultural Management*, 3: 27–44

Ratha, Dilip and Xu, Zhimei (2008), *Migration and Remittances Factbook*, World Bank, available online at: http://econ.worldbank.org/WBSITE/EXTERNAL/EXTDEC/EXTDECPROSPECTS/0,,contentMDK:21352016~menuPK:3145470~pagePK:64165401~piPK:64165026~theSitePK:476883,00.html

Regets, M. (2001), *Research and Policy Issues in High-Skilled International Migration: A Perspective with Data from the United States*, National Science Foundation, Discussion Paper No. 366, Arlington and IZA, Bonn, September

Reich, R. (1991), *The Work of Nations: Preparing Ourselves for 21st-Century Capitalism*, London: Simon and Schuster

Robinson, V. and Carey, M. (2000), 'Peopling skilled international migration: Indian doctors in the UK', *International Migration*, 38: 89–108

Rogers, E. M. (2003), *Diffusion of Innovations*, fifth edition, New York: Free Press (first edition 1962)

Romer, P. M. (1990), 'Endogenous technological change', *Journal of Political Economy*, 98(5): S71–S102

Romer, P. (1993), 'Idea gaps and object gaps in economic development', *Journal of Monetary Economics*, 32: 543–73

Ruhs, M. (2005), 'The potential of temporary migration programmes in future international migration policy', paper prepared for the Policy Analysis and Research Programme of the Global Commission on International Migration, Oxford

Ruhs, M. and Anderson, B. (2006), *Semi Compliance in the Migrant Labour Market*, Oxford: University of Oxford, Centre on Migration, Policy and Society, Working Paper 30

Ruhs, M., Anderson, B., Rogaly, B. and Spencer, S. (2006), *Changing Status, Changing Lives? Methods, Participants and Lessons Learnt*, Oxford: University of Oxford, Centre on Migration, Policy and Society

Sakai, J. (2000), *The Clash of Economic Cultures: Japanese Bankers in the City of London*, New Brunswick, NJ: Transaction Publishers

Salt, J. (1992), 'Migration processes among the highly skilled in Europe', *International Migration Review*, 26: 484–505

Salt, J. (1997), *International Movements of the Highly Skilled*, Paris: OECD, International Migration Unit Occasional Papers No. 3

Salt, J. (2003), *Current Trends in International Migration in Europe*, Strasbourg: Council of Europe

Salt, J. and Findlay, A. (1989), 'International migration of highly skilled manpower: theoretical and development issues', in Appleyard, R. (ed.), *The Impact of International Migration on Developing Countries*, Paris: OECD, pp. 109–28

Salt, J. and Ford, R. (1993), 'Skilled international migration in Europe: the shape of things to come?', in King, R. (ed.), *Mass Migration in Europe*, London: Belhaven Press, pp. 293–309

Sassen, S. (1998), *Globalization and its Discontents*, New York: New Press

Sassen, S. (2000a), *Cities in a World Economy*, second edition, Thousand Oaks, CA: Pine Forge Press

Sassen, S. (2000b), 'Regulating immigration in a global age: a new policy landscape', *Annals of the American Academy*, 570: 65–77

Saxenian, A. L. (1999), *Silicon Valley's New Immigrant Entrepreneurs*, San Francisco: Public Policy Institute of California

Saxenian, A. L. (2002), 'Silicon Valley's new immigrant high-growth entrepreneurs', *Economic Development Quarterly*, 16: 20–31

Saxenian, A. L. (2006), *The New Argonauts*, Cambridge, MA: Harvard University Press

Saxenian, A. L. and Hsu, J.-Y. (2001), The Silicon Valley–Hsinchu connection: 'technical communities and industrial upgrading', *Industrial and Corporate Change*, 10: 893–920

Schoenberger, E. (1997), *The Cultural Crisis of the Firm*, Oxford: Blackwell

Schrecker, T. and Labonte, B. (2004), 'Taming the brain drain: a challenge for public health systems in Southern Africa', *International Journal of Occupational and Environmental Health*, 10: 409–15

Scullion, H. (2001), 'International human resource management', in Storey, J. (ed.), *Human Resource Management*, London: International Thomson

Sennett, R. (2000), *The Corrosion of Character: The Personal Consequences of Work in the New Capitalism*, London: W. W. Norton and Company

Shields, M. A. and Price, S. W. (2003), *The Labour Market Outcomes and Psychological Well-being of Ethnic Minority Migrants in Britain*, London: Home Office, Home Office Online Report 07/03

Simmie, J. (2003), 'Innovation and urban regions as national and international nodes for the transfer and sharing of knowledge', *Regional Studies*, 37: 607–20

Sjaastad, L. A. (1962), 'The costs and returns of human migration', *Journal of Political Economy*, 70: 80–93

Smith, C. (2006), 'The double indeterminacy of labour power: labour effort and labour mobility', *Work, Employment and Society*, 20: 389–42

Smith, H. L. and Waters, R. (2005), 'Employment mobility in high-technology agglomerations: the cases of Oxfordshire and Cambridgeshire', *Area*, 37: 189–98

Smith, M. P. (2001), *Transnational Urbanism: Locating Globalization*, Oxford: Blackwell

Smith, M. P. (2005), 'Transnational urbanism revisited', *Journal of Ethnic and Migration Studies*, 31: 235–44

Solow, R. (1956), 'A contribution to the theory of economic growth', *Quarterly Journal of Economics*, 70: 65–94

Solow, R. (1957), 'Technical change and the aggregate production function', *Review of Economics and Statistics*, 39: 312–20

Song, J., Almeida, P. and Wu, G. (2003), 'Learning-by-hiring: when is mobility more likely to facilitate interfirm knowledge transfer?', *Management Science*, 49: 351–65

Soto, M. (2002), *Rediscovering Education in Growth Regression*, Paris: OECD, Technical Paper 202, OECD Development Centre, Empowering People to Meet the Challenges of Globalization Research Programme

Stark, O. (1991), *The Migration of Labour*, Oxford: Blackwell

Stark, O. (2003), *Rethinking the Brain Drain*, Calgary, AB: Discussion Paper 2003–04, Department of Economics, University of Calgary

Sternberg, R. (1995), 'Theory and management of tacit knowledge as a part of practical intelligence', *Zeitschrift für Psychologie*, 203: 319–34

Storper, M. and Venables, A. J. (2002), 'Buzz: the economic force of the city', paper presented at the DRUID Summer Conference on 'Industrial Dynamics of the New and Old Economy – Who Is Embracing Whom?', Elsinore, Copenhagen, 6–8 June. Quoted in Bathelt *et al.* 2004

Straubhaar, T. (2000), *International Mobility of the Highly Skilled: Brain Gain, Brain Drain or Brain Exchange*, Hamburg: Hamburg Institute of International Economics, Discussion Paper 88

Styhre, A. (2004), 'Rethinking knowledge: a Bergsonian critique of the notion of tacit knowledge', *British Journal of Management*, 15: 177–88

Subramanian, A. and Roy, D. (2001), *Who Can Explain the Mauritian Miracle: Meade, Romer, Sachs, or Rodrik?*, Washington, DC: International Monetary Fund, Working Paper 01/11

Tarrius, A. (1996), 'Territoires circulatoires des migrants et espaces européens', in Berthelot, J.-M. and Hirschorn, M. (eds), *Mobilités et Ancrages*, Paris: L'Harmattan

Taylor, J. E. (1986), 'Differential migration, networks, information and risk', in Stark, O. (ed.), *Research in Human Capital and Development*, vol. 4: *Migration, Human Capital, and Development*, Greenwich, CT: JAI Press, pp. 147–71

Taylor, J. E. (1987), 'Undocumented Mexico–U.S. migration and the returns to households in rural Mexico', *American Journal of Agricultural Economics*, 69: 16–38

Taylor, K. (2002), 'Is imagination more important than knowledge?', *Times Higher Education Supplement*, 20 December

Taylor, S. and Osland, J. S. (2003), 'The impact of intercultural communication on global learning', in Easterby-Smith, M. and Lyles, M. A. (eds), *The Blackwell Handbook of Organizational Learning and Knowledge*, Oxford: Blackwell, pp. 212–32

Thomas-Hunt, M. C., Ogden, T. Y. and Neale, M. A. (2003), 'Who's really sharing? Effects of social and expert status on knowledge exchange within groups', *Management Science*, 49: 464–77

Thomi, W. and Böhn, T. (2003), *Knowledge Intensive Business Services in Regional Systems of Innovation – Initial Results from the Case of Southeast Finland*, 43rd European Congress of the Regional Science Association, Jyväskylä, Finland, 27–30 August

Thompson, P., Warhurst, C. and Callaghan, G. (2001), 'Ignorant theory and knowledgeable workers: interrogating the connections between knowledge, skills and service', *Journal of Management Studies*, 38: 923–42

Tomlinson, F. and Egan, S. (2002), 'From marginalization to (dis)empowerment: organizing training and employment services for refugees', *Human Relations*, 55: 1019–43

Tung, R. L. (1998), 'American expatriates abroad: from neophytes to cosmopolitans', *Journal of World Business*, 33: 125–44

Tushman, M. and Scanlan, T. (1981), 'Boundary spanning individuals: their role in information transfer and their antecedents', *Academy of Management Journal*, 24: 289–305

United Nations (2002), *World Population Prospects: The 2002 Revision, Vol. I: Comprehensive Tables*, New York: United Nations, Department of Economic and Social Affairs/Population Division

United Nations (2003), *International Migration 2002*, New York: United Nations, Department of Economic and Social Affairs/Population Division (ST/ESA/SER.A/219)

United Nations (2004a), *World Economic and Social Survey 2004: International Migration*, New York: United Nations, Department of Economic and Social Affairs

United Nations (2004b), *World Population Prospects: The 2004 Revision, Vol. III: Analytical Report*, New York: United Nations, Department of Economic and Social Affairs/Population Division

UN Population Division (2005), *Trends in Total Migrant Stock: The 2005 Revision*, New York: UN Population Division of the Department of Economic and Social

Affairs of the United Nations Secretariat, http://www.un.org/esa/population/ publications/migration/UN_Migrant_Stock_Documentation_2005.pdf (accessed 8 February 2008)

Urry, J. (2000), *Sociology beyond Societies: Mobilities for the Twenty-First Century*, London: Routledge

Vertovec, S. (2002), *Transnational Networks and Skilled Labour Migration*, Oxford: University of Oxford, ESRC Transnational Communities Programme, Working Paper Series, WPTC-02-02

Vertovec, S. (2004), *Trends and Impacts of Migrant Transnationalism*, Oxford: University of Oxford, Centre on Migration Policy and Society, Working Paper No. 3

Voydanoff, P. (2001), 'Incorporating community into work and family research: a review of basic relationships', *Human Relations*, 54: 1609–37

Waldinger, R., Aldrich, H. and Ward, R. (1990), 'Opportunities, group character- istics, and strategies', in Waldinger, R., Aldrich, H. and Ward, R. (eds), *Ethnic Entrepreneurs*, Sage, Newbury Park, CA, pp. 13–48

Waldinger, R. and Lichter, M. I. (2003), *How the Other Half Works: Immigration and the Social Organization of Labour*, Berkeley, CA: University of California Press

Ward, B. and Glenn, J. (2005), *Dr. Space: The Life of Werner von Braun*, Annapolis, MD: Naval Institute Press

Ward, K. (2004), 'Going global? Internationalisation and diversification in the tem- porary staffing industry', *Journal of Economic Geography*, 4: 251–73

Ward, S. (2005), 'Low-cost airlines boost Scots business boom', *The Scotsman*, 1 June

Warhurst, C. and Thompson, P. (1998), 'Hands, hearts and minds: changing work and workers at the end of the century', in Thompson, P. and Warhurst, C. (eds), *Workplaces of the Future*, London: Macmillan, pp. 1–24

Warnes, A. M. and Williams, A. M. (2006), 'Older migrants in Europe: an innovative focus for migration studies', *Journal of Ethnic and Migration Studies*, 32: 1257–81

Welch, J. (2001), 'Timeless principles', *Executive Excellence*, February, p. 21

Wenger, E. (1998), *Communities of Practice: Learning, Meaning, and Identity*, Cambridge: Cambridge University Press

Wenger, E. (2000), 'Communities of practice and social learning systems', *Organizations*, 7: 225–46

Westwood, R. and Low, D. R. (2003), 'The multicultural muse; culture, creativity and innovation', *International Journal of Cross Cultural Management*, 3: 235–59

Williams, A. (2001), 'New forms of international migration: in search of which Europe', in Wallace, H. (ed.), *Interlocking Dimensions of European Integration*, Basingstoke: Palgrave, pp. 103–21

Williams, A. M. (2005), *International Migration and Knowledge*, Oxford: University of Oxford, Centre on Migration, Policy and Society, Working Paper 17

Williams, A. M. (2006a), 'Lost in translation? International migration, learning and knowledge', *Progress in Human Geography*, 30: 588–607

Williams, A. M. (2006b), 'Space and mobility in the knowledge-based economy', paper presented at the Knowledge-based economy conference, at the University of Lancaster, 31 August–2 September

Williams, A. M. (2006c), 'International migration and knowledge: an anti-elitist agenda', *Ad-Lib: Journal for Continuing Liberal Adult Education*, 30: 2–3

Williams, A. M. (2007a), 'Listen to me, learn with me: International migration and knowledge', *British Journal of Industrial Relations*, 45: 361–82

Williams, A. M. (2007b), 'International labour migration and tacit knowledge trans-actions: a multi-level perspective', *Global Networks* 7: 1–22

Williams, A. M. and Baláž V. (2004), 'From private to public sphere, the com-modification of the au pair experience? Returned migrants from Slovakia to the UK', *Environment and Planning* A 36: 1813–33

Williams, A. M. and Baláž, V. (2005a), 'What human capital, which migrants? Returned skilled migration to Slovakia from the UK', *International Migration Review* 39: 439–68

Williams, A. M. and Baláž, V. (2005b), 'Winning then losing the battle with global-ization: Vietnamese petty traders in Slovakia', *International Journal of Urban and Regional Research*, 29: 533–49

Williams, A. M. and Baláž, V. (2008a), 'Low cost carriers, economies of flow, and regional externalities', *Regional Studies*, in press

Williams, A. M. and Baláž (2008b), *International Return Mobility, Learning and Knowledge Transfer: Slovak Doctors*, London: London Metropolitan University, Institute for the Study of European Studies, Working Paper 1

Williams, A. M., Balaz, V. and Wallace, C. (2004), 'International labour mobility and uneven regional development in Europe: human capital, knowledge and entrepreneurship', *European Urban and Regional Studies*, 11: 27–46

Williams, A. M. and Hall, C. M. (2002), 'Tourism, migration, circulation and mobil-ity: the contingencies of time and place', in Hall, C. M. and Williams, A. M. (eds), *Tourism and Migration: New Relationships between Production and Consumption*, Dordrecht: Kluwer, pp. 1–52

Winkelmann-Gleed, A. (2006), *Migrant Nurses: Motivation, Integration and Contribution*, Abingdon, Oxfordshire: Radcliffe Publishing

World Bank (1999), *World Development Report 1999: Knowledge for Development*, New York: Oxford University Press

Wu, B. and Morris, J. (2006), ' "A life on the ocean wave": the "postsocialist" careers of Chinese, Russian and Eastern European seafarers', *International Journal of Human Resource Management*, 17: 25–48

Yang, B. (2003), 'Toward a holistic theory of knowledge and adult learning', *Human Resource Development Review*, 2: 106–29

Yeoh, B. S. A. and Willis, K. (2005), 'Singaporean and British transmigrants in China and the cultural politics of "contact zones"', *Journal of Ethnic and Migration Studies*, 31: 269–85

Zhou, Y. and Tseng, Y-F. (2002), 'Regrouping the "ungrounded empires": localization as the geographical catalysts for transnationalism', *Global Networks*, 1: 131–53

Zook, M. (2004), 'The knowledge brokers: venture capitalists, tacit knowledge and regional development', *International Journal of Urban and Regional Research*, 28: 621–41

Zuboff, S. (1988), *In the Age of the Smart Machine: the Future of Work and Power*, New York: Basic Books

Zweig, D. (2006), 'Learning to compete: China's efforts to encourage a "reverse brain drain" ', in Kuptsch, C. and Fong, P. E. (eds), *Competing for Global Talent*, Geneva: International Labour Organization, pp. 187–213

Zweig, D. and Changgui, C. (1995), *China's Brain Drain: Overseas Chinese Students and Scholars in the 1990s*, Berkeley, CA: University of California, Institute of East Asian Studies

Zweig, D., Rosen, S. and Changgui, C. (2004), 'Globalization and transnational human capital: overseas and returnee scholars to China', *China Quarterly*, 179: 735–57

Index

eBooks – at www.eBookstore.tandf.co.uk

A library at your fingertips!

eBooks are electronic versions of printed books. You can store them on your PC/laptop or browse them online.

They have advantages for anyone needing rapid access to a wide variety of published, copyright information.

eBooks can help your research by enabling you to bookmark chapters, annotate text and use instant searches to find specific words or phrases. Several eBook files would fit on even a small laptop or PDA.

NEW: Save money by eSubscribing: cheap, online access to any eBook for as long as you need it.

Annual subscription packages

We now offer special low-cost bulk subscriptions to packages of eBooks in certain subject areas. These are available to libraries or to individuals.

For more information please contact webmaster.ebooks@tandf.co.uk

We're continually developing the eBook concept, so keep up to date by visiting the website.

www.eBookstore.tandf.co.uk

For Product Safety Concerns and Information please contact our EU
representative GPSR@taylorandfrancis.com
Taylor & Francis Verlag GmbH, Kaufingerstraße 24, 80331 München, Germany

www.ingramcontent.com/pod-product-compliance
Lightning Source LLC
Chambersburg PA
CBHW050427280326
41932CB00013BA/2026

9 780415 761840